S0-BRB-334

Rhubarb Recipes

Rhubarb Recipes

Second Edition

Compiled by Jeanne DeMars

APPLE
BLOSSOM
BOOKS

© 1994 Jeanne DeMars
ISBN 0-9627766-2-9
Second Edition 1994

Rhubarb Illustrations
Tammy Pickar

Design & Typography
Bill Balcziak
The Final Word

Publisher
Apple Blossom Books
P.O. Box 134
Belle Plaine, Minnesota 56011
U.S.A.
(612) 873-6620

How to Make Rhubarb Wine

by Ted Kooser

Go to the patch some afternoon
in early summer, fuzzy with beer
and sunlight, and pick a sack
of rhubarb (red or green will do)
and God knows watch for rattlesnakes
or better, listen; they make a sound
like an old lawn mower rolled downhill.
Wear a hat. A straw hat's best
for the heat but lets the gnats in.
Bunch up the stalks and chop the leaves off
with a buck knife and be careful.
You need ten pounds; a grocery bag
packed full will do it. Then go home
and sit barefooted in the shade
behind the house with a can of beer.
Spread out the rhubarb in the grass
and wash it with cold water
from the garden hose, washing
your feet as well. Then take a nap.
That evening, dice the rhubarb up
and put it in a crock. Then pour
eight quarts of boiling water in,
cover it up with a checkered cloth
to keep the fruit flies out of it,
and let it stand five days or so.
Take time each day to think of it.

Ferment ten days, under the cloth,
sniffing of it from time to time,
then siphon it off, swallowing some,
and bottle it. Sit back and watch
the liquid clear to honey yellow,
bottled and ready for the years,
and smile. You've done it awfully well.

From Sure Signs: Selected Poems,
University of Pittsburgh Press, 1980

Table of Contents

Acknowledgement

As in my first edition, I want to thank all my friends and relatives (and their friends and relatives) who generously shared their rhubarb recipes with me. All the old favorite recipes are still present in this new edition, but due to the wonderful response from those who tried the recipes in my first book, I searched far and wide for even more ways to prepare rhubarb. Rhubarb is more versatile than I ever imagined. As a result, I have added more than 120 new recipes and rearranged and added categories. So, thanks to those of you who have written to me with your comments, and to Terry Kerwin of the International Rhubarb Society, as well as Dale Marshall, agricultural engineer at Michigan State University.

Jeanne DeMars

Planting & Growing

HISTORY

Rhubarb, or pie plant, as it is sometimes called, is a hardy cool-weather perennial with broad dark-green leafy tops and green or red stalks. It flourishes in the northern half of the United States where temperatures drop below 50° in the winter. The cold temperatures are needed to break the dormancy of the plant.

Although used as a fruit, it is actually part of the vegetable family. The plant seems to have come to us from Asia, probably Siberia or northern Tibet, then into Europe by the 14th century. It was introduced to Britain around 1578. It was grown for its medicinal (mild laxative) effect and ornamental properties until the 19th century, when it came into use for general baking and cooking.

VARIETIES & GROWING CONDITIONS

There are several varieties of rhubarb such as Crimson Wine, Valentine, McDonald, and Ruby, which produce red stalks and have coarse deep-green foliage. According to a USDA pamphlet on rhubarb production, rhubarb grows well in almost any type of soil that is well-drained and fertile. Liberal quantities of fertilizer are helpful. An application of manure mixed with straw in the early spring will both feed the plant and act as a mulch to keep moisture in the ground and to control weeds. Otherwise a 10-10-10 fertilizer worked in around the plant in the spring will also do.

PLANTING

To plant rhubarb the USDA recommends dividing crowns of existing plants rather than planting seeds. Dig up dormant crowns that are at least two to three years old and split between the large buds. Try to leave on as large a piece of storage root as possible with each bud. Protect the pieces from drying out and plant as soon as possible about 2-3 feet apart, covering with 2-3 inches of soil pressed firmly around each piece.

DISEASES

Rhubarb is susceptible to very few problems, according to the USDA. One of the possible problems is a disease called foot rot which causes lesions at the base of the stalk, eventually killing the entire plant. There is no remedy for this disease. The other problem is an insect called the rhubarb curculio, a rusty snout beetle about ¾-inches long. It bores into stalks, crowns, and roots, destroying the plant. It also attacks wild dock, a weed which closely resembles rhubarb. The USDA recommends burning infested plants and destroying wild dock growing near the rhubarb after the beetles have laid their eggs in July.

HARVESTING

Rhubarb can be harvested when the stalks are about 12-15 inches in length and about the size of a thumb's thickness, sometime in early spring. Usually you should harvest for no more than two months to insure a hardy plant for the next year. Harvest the largest and best stalks by holding on near the base of the plant and pulling them to one side. Do not cut the rhubarb stalk. Thinning out the smaller stalks will help development of the remaining ones. When seedstalks appear, cut them off. Then the plant will not be putting energy into making seeds but into the leaves and stalks of the plant. The USDA says a heavy crop of rhubarb depends on strong leaf growth the year before.

WARNING

Only the stalk portion of the rhubarb plant is edible. ***Do not eat the leaves,*** cooked or uncooked, since they contain amounts of oxalic acid in a quantity large enough to be poisonous.

Because of the acid, boiling the leaves can be a natural way to clean pans. You may have noticed that even boiling the stalks, when cooking a sauce, will make your pans shine.

NUTRITIONAL COMPOSITION

Rhubarb is almost 95% water, containing vitamins A and C, potassium, and carbohydrates. It contains calcium but in a form your body can't utilize because of the acid in the plant. There is no fat or cholesterol and very few calories. Despite the stringy nature of the stalks, there is very little fiber in rhubarb.

Baking Tips

The acid in the rhubarb can cause a reaction in a metal baking dish which gives a metallic undertaste to the baked product, so use glass or teflon-coated bakeware if you have it.

When using frozen rhubarb in recipes written for fresh, add a little more thickening agent to the recipe. More of the water content of the rhubarb seems to cook out when using it frozen.

Rhubarb makes beautiful moist cakes, breads, and muffins but don't store baked products in airtight containers because they will get soggy.

Preserving Rhubarb

DRYING

To prepare for drying, cut the leaves off the stalks and discard. Wash and cut the stalks into 1-inch chunks, steam for 1-2 minutes until slightly tender, then dry at 150° for 2-3 hours until tough to crisp.

FREEZING

Cut the leaves off the stalks and discard. Wash and cut the stalks into 1- to 2-inch pieces. Freeze pieces on trays or cookie sheets, then transfer to plastic freezer bags or ice cream buckets. Rhubarb does not have to be blanched. If you prefer, you can blanch for one minute and cool immediately before freezing. It is also optional to add sugar at about ½ cup for each quart frozen in containers. Leave a ½-inch head-space in the container.

CANNING

Boiling water method: Cut the leaves off the stalks and discard. Wash and cut the stalks into ½-inch pieces. Add ½ to 1 cup of sugar to 1 quart of rhubarb. Let the rhubarb and sugar stand in a cool place for three or four hours. Slowly bring to a boil. Boil no more than a minute or the pieces will break up. Pack hot into hot jars, leaving ½-inch head-space. Adjust lids and process in boiling water bath for 10 minutes for either pints or quarts. Remove jars and complete seals if necessary.

For rhubarb juice: Cut the leaves off the stalks and discard. Wash and cut the stalks into pieces no longer than 4 inches. Add 1 quart of water to 4 quarts (5 pounds) of rhubarb and bring just to boiling. Press through a jelly bag, then re-strain through damp cheesecloth. Sweeten to taste with about ½ cup sugar per quart of juice.

To can: Reheat juice to simmering and pour into hot sterilized jars, leaving ½-inch headspace. Process in boiling water bath (pints or quarts) for 10 minutes.

Fruit Salads and Ambrosias

Lemon Rhubarb Gelatin Salad
Jellied Rhubarb Mold
Rhubarb Cream Cheese Salad
Rhubarb Salad Supreme
Pineapple-Rhubarb Salad
Rhubarb Delight Salad
Pineapple Rhubarb Ring
Rhubarb-Apple Fruit Salad
Rhubarb 24-hour Salad
Mandarin Rhubarb Salad
Rhubarb Molded Salad
Tart Rhubarb Salad
Strawberry-Rhubarb Ring
Cranberry Rhubarb Salad Squares
Fresh Bananas With Ruby Sauce
Rhubarb Ambrosia
Rhubarb-Raspberry Salad
Rhubarb Vegetable Salad

Lemon Rhubarb Gelatin Salad

    1 envelope unflavored gelatin
    ¼ cup lemon juice
    ½ cup sugar
    2 cups cooked rhubarb, slightly
      sweetened and cold
    ¼ cup cold water
    1 cup boiling water

Cover gelatin with ¼ cup cold water and let gelatin soften (about two minutes). Add lemon juice, 1 cup boiling water and sugar. Stir until thoroughly dissolved. Cool mixture, then place in refrigerator. When it starts to set, add cold rhubarb and mix well. Return to refrigerator until set. Makes four cups (8 servings). This makes a lovely soft salad.

Jellied Rhubarb Mold

    1½ pounds tender young rhubarb
    ¾-1 cup sugar
    ½ cup boiling water
    3 teaspoons gelatin
    1½ tablespoons cold water
    ½ cup heavy cream, whipped
      Grated rind of ½ orange

Wash but do not peel the rhubarb, dice, then place in a heavy saucepan with the sugar and water. Cover and cook over a low flame about 15 minutes, stirring occasionally. Drain off the hot juice and add to gelatin, previously softened in the cold water. Stir until dissolved, then pour over the rhubarb itself which has been divided among individual serving glasses. When cold, chill in the refrigerator and serve topped with a puff of slightly sweetened whipped cream flavored with the grated orange rind. Serves 4 to 6.

Rhubarb Cream Cheese Salad

    1 package strawberry or lemon
      gelatin
    2 cups cooked and sweetened
      rhubarb
    2 tablespoons orange concentrate
      or juice of 1 orange
      Juice of 1 lemon
    1 package cream cheese
    ½ cup chilled and whipped cream

Dissolve gelatin in sweetened rhubarb. Add orange and lemon juice; chill until partially set. Blend in cream cheese. Fold in whipped cream. Chill overnight. If lemon gelatin is used, a few drops of red food coloring may be added. Makes 8 servings.

Rhubarb Salad Supreme

    ¾ cup rhubarb
    1 cup cream (whipped)
    1 3-oz. package cream cheese
    ½ cup chopped nutmeats
    1 cup miniature marshmallows
    1 package cherry gelatin

In saucepan, slowly cook rhubarb in small amount of water until tender. Set aside. Dissolve gelatin in one cup of hot water. Cool until syrupy. Whip cream and cream cheese together until stiff. Add gelatin and beat well. Add other ingredients. Stir until it begins to set. Refrigerate four hours. Serve as light lunch with rye crisps and relish tray.

Pineapple-Rhubarb Salad

1 cup cooked rhubarb
1 package lemon gelatin
1 cup crushed pineapple
1½ cups liquid

Drain the juice from the fruits for some of the liquid and add water to equal 1½ cups. Heat; dissolve gelatin in the heated juice. Chill slightly before adding rhubarb and pineapple. Place in refrigerator to set. Makes 4-6 servings. (Your rhubarb will be sweeter if you cook it until done, and then add sugar to taste.)

Rhubarb Delight Salad

1 package raspberry gelatin
1½ cups liquid
1 cup marshmallows
2 cups cooked rhubarb
1 cup crushed pineapple
½ cup Cool Whip

Drain the juice from the fruits for some of the liquid and add water to equal 1½ cups. Heat and dissolve gelatin in the heated juices. Add fruits after the gelatin is partially set and whipped.

Pineapple Rhubarb Ring

2½ cups pineapple tidbits
2 cups fresh rhubarb, cut in 1" pieces
⅓ cup sugar
½ cup broken pecans
½ cup water
2 3-oz. packages cherry gelatin
1 tablespoon lemon juice

Drain pineapple, reserving syrup. Combine rhubarb, sugar and water. Cover and cook until just tender (about five minutes). Drain thoroughly, reserving syrup. Combine pineapple and rhubarb syrups; add water to make about 3 cups. Heat to boiling, add gelatin and stir to dissolve. Add lemon juice. Cool. Chill until partially set. Fold in rhubarb, pineapple and nuts. Pour into mold and chill until firm. Unmold on greens and serve with mayonnaise. Makes 8-10 servings.

Rhubarb-Apple Fruit Salad

2 cups rhubarb, finely cut
½ cup sugar
½ cup water

Combine rhubarb, sugar and water in saucepan and cook, covered, until soft, 10 minutes or less. Cool.

1 3-oz. box strawberry gelatin
1 cup boiling water
½ cup sugar

Dissolve gelatin in boiling water. Stir until dissolved. Chill, then add to rhubarb.

½ cup well drained crushed pineapple
2 cups finely chopped apples
½ cup finely chopped nutmeats

Add pineapple, apples and nuts to rhubarb mixture. Chill until set (several hours). Note: the mixture may be poured into an oiled 8" square pan.

Rhubarb 24-hour Salad

2 cups diced rhubarb
2 cups diced pineapple
2 cups miniature marshmallows
¼ pound walnuts
2 eggs
2 tablespoons sugar
¼ cup light cream
   juice of one lemon
1 cup heavy cream (whipped)

Combine pineapple and rhubarb. Add and set aside marshmallows and walnuts. Beat eggs till light. Gradually add and mix sugar, light cream, lemon juice and rhubarb. Cook in double boiler till thick and smooth, stirring constantly. Cool and fold in the whipped cream. Pour over fruit mixture and mix lightly. Chill for 24 hours. Do not freeze.

Mandarin Rhubarb Salad

4 cups diced fresh rhubarb
1 cup water
¾ cup sugar
¼ teaspoon salt
2 eleven-ounce cans mandarin
   oranges
2 3-oz. packages strawberry gelatin
1¾ cup cold water
¼ cups lemon juice
   Dairy sour cream
   Nutmeg

Combine rhubarb, water, sugar and salt. Bring to a boil. Reduce heat and simmer only until rhubarb loses crispness. Remove from heat and add gelatin. Stir until dissolved. Add cold water and lemon juice. Chill until partially thickened. Spoon into 8"x8" pan or eight individual molds. Chill until firm. Serve on lettuce, top with sour cream and sprinkle with nutmeg.

Rhubarb Molded Salad

1 package frozen or 2 cups fresh
   rhubarb
2 packages strawberry gelatin
2 cups pineapple juice
2 cups apples, chopped
1 cup nuts, chopped
   Pinch of salt

Cook rhubarb as directed on package. Add 1 cup pineapple juice; bring to a boil. Add gelatin and stir until dissolved. Add salt and remaining pineapple juice. Let cool. Add apples and nuts. Chill until firm.

Tart Rhubarb Salad

4   cups diced fresh rhubarb
¾   cup sugar
2   packages strawberry gelatin
¼   cup lemon juice
2   cans mandarin oranges, drained
1   cup water
¼   teaspoon salt
1½ cups cold water
1   cup diced celery

Combine rhubarb, water, sugar and salt. Bring to a boil. Reduce heat and simmer only until rhubarb loses its crispness. Remove from heat; add gelatin. Add cold water and lemon juice. Chill until partially set. Fold in oranges and celery. Put in pan. Serve on lettuce with a little sour cream and dash of nutmeg.

Strawberry-Rhubarb Ring

3 cups fresh rhubarb, cut in small
    pieces
2 cups sugar
1 cup boiling water
2 envelopes gelatin
½ cup cold water
    Strawberries
    Powdered sugar
1 cup heavy cream, whipped
1 tablespoon sugar
2 tablespoons kirsch

Cook the rhubarb with the sugar and boiling water in a two-quart saucepan until tender. Dissolve the gelatin in cold water, add to the rhubarb, and stir until fully dissolved. Pour into an eight-cup ring mold which has been oiled. Chill until firm. To serve, unmold the ring onto a serving plate, fill the center with strawberries and dust with powdered sugar. Serve with whipped cream, into which the sugar and kirsch has been folded.

Cranberry Rhubarb Salad Squares

1 package black cherry gelatin
¾ cup very hot water
1 can (one pound) whole cranberry
    sauce
¼ cup ginger ale
    sour cream
    cinnamon
¼ cup finely diced rhubarb

Dissolve gelatin in hot water. Stir in cranberry sauce, rhubarb and ginger ale. Pour into an 8"x8" pan. Chill until firm. Cut into squares and serve with lettuce garnish. Top with one tablespoon of sour cream and a dash of cinnamon. Makes 5-6 salads. To make squares more like a dessert, pour into dessert glasses before chilling.

Fresh Bananas With Ruby Sauce

1 pound fresh rhubarb, cut into
    1-inch pieces
¼ cup fresh orange juice
¾-1 cup sugar
½ teaspoon grated fresh orange rind
¼ teaspoon cinnamon
    Dash salt
⅓ cup slivered almonds (optional)
6 bananas

In medium saucepan combine rhubarb, orange juice, sugar, lemon rind, cinnamon and salt. Bring to a boil, reduce heat and simmer, partially covered, about 10 minutes or until rhubarb begins to fall apart into strands, stirring occasionally. Chill. At serving time, peel bananas and cut in half lengthwise and crosswise. Place one banana on each of 6 dessert plates and top with rhubarb sauce. Sprinkle with almonds. Makes 6 servings.

Rhubarb Ambrosia

    ½ cup water
    ½ cup sugar
    3 cups rhubarb (1-inch pieces)
    2-3 oranges
    1 banana
    ½ cup flaked coconut

Mix sugar and water in a saucepan. Bring mixture to a boil and then stir in rhubarb. Cover, bring to a boil again, and turn off heat. Allow to stand covered until room temperature. Refrigerate until cold. Remove peel from oranges and slice. Arrange orange slices overlapping around the edge of a serving dish. Slice banana and add to rhubarb. Spoon rhubarb, banana slices and juice, if desired, into center of the oranges and sprinkle coconut over all. Serves 6.

Rhubarb-Raspberry Salad

    3 cups cut-up rhubarb
    1 cup water
    1 3-oz. package raspberry-flavored
      gelatin
    1 cup sugar
    1 cup diced celery
    1 cup nuts

Cook rhubarb in the water (about 10 minutes, until rhubarb is tender). Add gelatin and sugar and stir until dissolved. Cool until syrupy. Add celery and nuts. Chill until firm (do not try to mold this salad, because it does not get that firm). For a dessert, substitute 1 cup miniature marshmallows for celery.

Rhubarb Vegetable Salad

    3 cups fresh rhubarb, cut in small
      pieces
    1¾ cups water
    ½ cup sugar
    1 3-oz. package lime gelatin
    ¼ teaspoon salt
    1 tablespoon green onions,
      chopped (optional)
    2 tablespoons green pepper (cut
      up)
    ½ cup cabbage, finely shredded
    1 cup celery, chopped
    1 tablespoon pimentos (or red
      pepper), chopped

Cook rhubarb and water five minutes. Add sugar; stir in gelatin. Add salt and vegetables. Pour into oiled mold; chill until set. Serve with salad dressing.

Relishes, Chutneys, Salsas & Meatsauces

Rhubarb and Onion Relish

Victoria Sauce

Barbecued Rainbow Trout with Rhubarb Relish

Spiced Rhubarb

Spiced Rhubarb for Meat

Rhubarb Chutney

Raisin-Rhubarb Chutney

Honey-Apricot-Rhubarb
Chutney

Rhubarb-Apple Chutney

Pineapple-Rhubarb Chutney

Sausage With Rhubarb Salsa

Rhubarb Salsa

Rhubarb BBQ Sauce

Rhubarbecue Sauce

Rhubarb Glaze for Chicken

Zesty Meat Sauce With Horseradish

Rhubarb and Onion Relish

8   cups sliced raw rhubarb
4   cups sliced onion
4   cups packed light brown sugar
2   teaspoons salt
1   teaspoon ground ginger
1   teaspoon ground cinnamon
1   cup vinegar
1   garlic bud, peeled and minced
1   tablespoon whole mixed pickling
    spices

In a large heavy kettle, combine all except the last two ingredients. Tie the garlic and pickling spices in a small cloth bag and add. Bring to boiling; reduce heat and simmer until rhubarb is just tender—about 10 minutes. Don't overcook. Remove the spice bag, ladle the hot relish into hot sterilized jars, adjust lids, and process in a boiling-water bath for 5 minutes to ensure the seal. About 3 pints.

Victoria Sauce

8   cups rhubarb, chopped in 1"
    pieces
½   cup chopped onion
1½ cups chopped raisins
3½ cups brown sugar
½   cup vinegar
1   teaspoon salt
1   teaspoon ginger
1   teaspoon cinnamon
1   teaspoon allspice

Slowly boil together the rhubarb, onion, raisins, sugar and vinegar until thick. Add spices about five minutes before removing sauce from heat. Pour boiling hot into sterilized jars. Seal immediately. Makes 7-8 one-half pints.

Barbecued Rainbow Trout with Rhubarb Relish

1   pint fresh strawberries
1   cup barbecue sauce
½   cup diced Vidalia or other sweet
    onion
3   ounces rice vinegar
½   teaspoon cracked black pepper
1   cup diced blanched rhubarb,
    stringy outside removed
4   ten-ounce rainbow trout fillets
    Olive oil nonstick cooking spray
5   medium strawberries, diced
    Lemon juice and zest to taste

Prepare charcoal for barbecuing. Hull 1 pint fresh strawberries and purée in food processor. Combine berry purée with barbecue sauce; set aside. Cook diced onions in vinegar briefly, until they release their juices. Add pepper and rhubarb to onions; cover and braise on low heat five minutes. Let cool to room temperature. Spray trout fillets well with olive oil spray; spray grill too. Char-grill fish-skin side down until crisp and trout is nearly done, then brush flesh side lightly with barbecue glaze and continue to grill briefly. Place fillets on warm serving plate. Combine cooled rhubarb mixture with diced strawberries and lemon juice and zest. Serve this rhubarb relish alongside trout. Makes four servings.

Spiced Rhubarb

2½ pounds rhubarb
2 pounds sugar
¾ cup vinegar
1 teaspoon cinnamon
½ teaspoon cloves

Wipe rhubarb, skin, and cut stalks in 1-inch pieces. Put in preserve kettle, add remaining ingredients, bring to boiling point. Simmer until the consistency of a marmalade. Fill jelly glasses with mixture, cool and seal.

Spiced Rhubarb for Meat

6 pounds rhubarb, cut into 1" cubes
4 pounds light brown sugar
1 pint vinegar
2 tablespoons cinnamon
2 scant tablespoons ground cloves

Combine ingredients and cook over moderate heat until thick. Pour immediately into sterile jars and seal.

Rhubarb Chutney

6 pounds rhubarb, sliced
1 pound onions, peeled and finely chopped
1 clove garlic, crushed
2 tablespoons mixed spice
1 tablespoon salt
4¼ cups vinegar
4¼ cups sugar

Place the rhubarb, onions, garlic, spice, salt and half the vinegar into a preserving pan. Bring to a boil, then reduce the heat and simmer until the rhubarb is very soft. Add sugar and remaining vinegar; simmer, stirring frequently, until the chutney is thick. Bottle, seal and label. Yields approximately 9 cups.

Raisin-Rhubarb Chutney

2 pounds rhubarb, finely chopped (about 4 cups)
2 lemons with peel, finely chopped
1 package (16 ounces) raisins
2 pounds brown sugar
¼ cup grated fresh gingerroot
2 cloves garlic, crushed

Combine all ingredients; simmer, uncovered, 2 to 2½ hours, stirring occasionally. Refrigerate. Chutney will keep in refrigerator 2 to 3 months. Makes 7 cups.

Honey-Apricot-Rhubarb Chutney

2 cups chopped rhubarb
2 cups chopped, dried apricot
1 small onion, minced
¼ teaspoon cayenne pepper
1 cup light honey
1 cup raisins
2½ cups cider vinegar
Grated rind and juice of one lemon
1 tablespoon grated fresh gingerroot
1 teaspoon ground allspice
½ teaspoon ground cloves
2 tablespoons ground cinnamon

Place all ingredients in a large, heavy-bottomed pot, and bring to a boil. Lower heat and simmer uncovered for 25 minutes, stirring occasionally to prevent scorching. Makes 7 cups.

Rhubarb-Apple Chutney

5 medium cooking apples
4 pounds rhubarb, chopped
4 medium onions, chopped
1 pound raisins
1 pound dark brown sugar
2½ cups malt vinegar
1 teaspoon cayenne pepper
1 teaspoon curry powder
1 teaspoon cinnamon
1 teaspoon ground ginger
½ teaspoon mace
½ teaspoon cloves

Place all ingredients in a large kettle. Cover and bring to a boil. Simmer slowly for about two hours until thick. Ladle into sterilized jars and seal. Makes about 7 pints.

Pineapple-Rhubarb Chutney

2 cups cider vinegar
2 pounds brown sugar
2 pounds chopped rhubarb
1 cup crushed pineapple, drained
1 pound raisins
½ cup chopped onion
2 tablespoons candied ginger
Grated rind of one lemon
¼ teaspoon cayenne pepper

Combine all ingredients and cook slowly 3-4 hours until thick. Pour into sterilized jars and seal. Makes about 6 cups.

Sausage With Rhubarb Salsa

½ pound mild Italian sausage or fresh bratwurst
1 cup apple juice
½ cup dry white wine
Rhubarb Salsa (recipe follows)

Pierce sausage several times. Place in skillet with apple juice and white wine. Cook over low to medium heat, turning occasionally until sausages swell up and are no longer pink in center, about 15 minutes. Meanwhile, prepare salsa. To serve, place half of sausages on each of two plates and spoon rhubarb salsa on side. Serve hot. Serves 2.

Rhubarb Salsa

1 tablespoon vegetable oil
1 tablespoon minced shallot (1 medium shallot)
1 tablespoon grated fresh gingerroot
½ cup chopped rhubarb (about 1 large stalk)
2 tablespoons vinegar
1 tablespoon sugar
Dash cayenne pepper
2 tablespoons dry white wine, dry vermouth or water

In medium saucepan, heat oil. Add shallot and gingerroot and sauté over low heat until shallots are tender, about 5 minutes. Add rhubarb, cider vinegar and sugar and simmer uncovered, 5 minutes. Cover and simmer 5 minutes longer until tender. Add cayenne and white wine and simmer another 5 minutes. Makes about ½ cup.

Rhubarb BBQ Sauce

½ pound rhubarb stalks, chopped
½ cup drippings *or* liquid from roasting meat
1 teaspoon chili powder
¼ teaspoon pepper
½ cup water
½ cup sugar
2 tablespoons ketchup
2-4 drops hot red pepper sauce

In pan, mix ½ cup meat drippings (fat removed), rhubarb, sugar, vinegar, ketchup and red pepper sauce. Heat and boil sauce five minutes. Stir often. Brush on meat; bake 15 minutes. Brush more on and bake until glazed and lightly browned, 15-20 minutes more. Serve with extra sauce.

Rhubarbecue Sauce

1 teaspoon chili powder
¼ teaspoon pepper
½ teaspoon salt
½ cup water
2 cups finely cut rhubarb
½ cup sugar
2 tablespoons cider vinegar
¼ cup ketchup
2-3 drops hot red pepper sauce
½ teaspoon garlic powder

Combine ingredients and cook slowly for five minutes. Brush over seasoned chicken, beef or ribs. Bake until almost done. Brush with sauce and bake 15 minutes longer. Serve the extra sauce with the meat.

Rhubarb Glaze for Chicken

Whip in blender:
1½ cups rhubarb sauce
Whip in blender:
½ cup water or pineapple juice
2 tablespoons salad oil
1 tablespoon lemon juice
¼ teaspoon ginger
¼ teaspoon dry mustard

Simmer 15-20 minutes. Brush over chicken during the last 15 minutes of roasting or broiling.

Zesty Meat Sauce With Horseradish

1 teaspoon prepared horseradish
2 teaspoons lemon juice
1 teaspoon grated lemon peel
¾ cup mayonnaise
1 teaspoon prepared mustard
1 cup slightly sweetened, partly liquid rhubarb sauce

Mix all ingredients together. Serve with chicken, on a boiled beef dinner, over stew. Makes 2 cups.

Jams & Jellies

Rhubarb Butter
Strawberry-Rhubarb Jam
Rhubarb-Apricot Jam
Dried Apricot-Rhubarb Jam
Blueberry-Rhubarb Jam
Rhubarb Surprise Jelly
Fresh Blueberry-Rhubarb Jam
Candy-Rhubarb Jam
Cherry Rhubarb Jelly
Rhubarb-Ginger Jam
Gooseberry-Rhubarb Jam
Rhubarb-Orange Jelly
Pineapple-Rhubarb Jam
Rhubarb-Raspberry Jam
Fresh Strawberry-Rhubarb Jelly
Sweet 'n Low Strawberry-Rhubarb Jam
Rhubarb Conserve
Pineapple-Rhubarb Conserve
Strawberry-Rhubarb Conserve
Rhubarb-Raisin Marmalade
Rhubarb-Carrot Marmalade
Rhubarb Fig Marmalade
Rhubarb-Zucchini Conserve
Rhubarb Marmalade

Rhubarb Butter

8 cups diced rhubarb
1 cup water
6 cups sugar
1 teaspoon cinnamon
½ teaspoon ground cloves

Place rhubarb and water in a kettle and simmer, stirring to prevent sticking, until rhubarb is very mushy and tender. Add the remaining ingredients and simmer, stirring often (about one-half hour) until very thick. Pour into hot sterilized jars and seal. Store in a cool, dark, dry place.

Strawberry-Rhubarb Jam

5 cups red rhubarb
4 cups sugar
1 3-oz. package strawberry gelatin

Cut rhubarb in small pieces. Add sugar and let stand overnight. In morning, boil 10 minutes. Add gelatin. Stir until dissolved. Put in jars and keep in refrigerator. Makes about 3 quarts.

Rhubarb-Apricot Jam

8 cups finely chopped rhubarb
4 cups sugar
1 can apricot pie filling
1 3-oz. package orange gelatin

Combine rhubarb and sugar in a bowl (not metal). Allow to stand overnight. In the morning, transfer to a pan, bring to a boil and simmer 10 minutes. Add the apricot pie filling and bring to a boil. Add the gelatin and stir until dissolved. Put in jars, cover and refrigerate or freeze.

Dried Apricot-Rhubarb Jam

8 ounces dried apricots
1½ cups water
4 cups chopped rhubarb
3 cups sugar

In large saucepan bring apricots and water to a boil. Simmer for 20 minutes. Add rhubarb and sugar. Simmer 40 minutes, stirring often. Ladle into sterilized jars. Seal.

Blueberry-Rhubarb Jam

5 cups rhubarb
1 cup water
5 cups sugar
1 can blueberry pie filling
2 3-oz. packages raspberry gelatin

Cook cut rhubarb in water until tender. Add sugar; cook about five minutes more, stirring constantly. Add pie filling and cook 6-8 minutes longer. Remove from heat. Add gelatin. Stir until completely dissolved. Pour in jars and seal. Store in refrigerator or freezer.

Rhubarb Surprise Jelly

7 cups rhubarb
3 cups sugar
1 can blueberry pie filling
1 large package raspberry gelatin

Mix first three ingredients; boil 10 to 15 minutes. Remove from heat; add gelatin and mix well. Pour into jars. Freeze or refrigerate. Number of servings depends on jar size.

Fresh Blueberry-Rhubarb Jam

3 cups rhubarb, cut small
2 cups fresh blueberries
2 cups sugar

Boil 10 minutes or until rhubarb is soft. Add 2 more cups of sugar and boil 5 minutes. Put in jars and seal with wax. For best consistency, keep jar in refrigerator after opening.

Candy-Rhubarb Jam

5 cups cut-up rhubarb
3 cups sugar
1 8-oz. jar maraschino cherries, cut up
1 10-oz. can crushed pineapple
1 one-pound package orange candy slices, cut up

Cook the rhubarb and sugar for 15 minutes; let stand overnight. Add the cherries, pineapple, and candy; boil for 15 minutes. Seal hot.

Cherry Rhubarb Jelly

4 cups cut-up rhubarb
4 cups sugar
1 can cherry pie filling
1 (3 ounce) box cherry gelatin

Mix the rhubarb and sugar and let set overnight. In the morning add pie filling and boil for 20 minutes. Take off the stove and add gelatin. Stir occasionally. Put back on low heat and pour into sterilized jars. This recipe can also be frozen.

Rhubarb-Ginger Jam

4 pounds rhubarb, washed, trimmed and cut into 1-inch pieces
1 cup water
Juice of 1 lemon
2-inch piece fresh root ginger, peeled and crushed
6 cups sugar
4 tablespoons crystallized ginger, finely chopped

Place the rhubarb, water and lemon juice in a preserving pan and bring to a boil. Add the root ginger, reduce the heat and simmer, stirring frequently, until the rhubarb is soft. Remove the piece of ginger. Add the sugar and stir until it has dissolved, add the crystallized ginger. Bring the mixture to a boil and boil rapidly for 10 to 15 minutes until setting point is reached. Bottle, cover and label. Approximate yield will be 18 cups.

Gooseberry-Rhubarb Jam

6 pounds gooseberries
4 pounds rhubarb, chopped
7 pounds sugar

Top and tail gooseberries. Place fruit in a large basin. Cover with sugar. Let stand overnight. Bring to a boil and simmer until fruit is tender and jam begins to set. Ladle into sterilized jars and seal.

Rhubarb-Orange Jelly

4 cups chopped uncooked rhubarb, not peeled
2⅓ cups water
1 (6 ounce) can frozen concentrated orange juice, thawed
1 (16 ounce) package powdered fruit pectin
4 cups sugar

Combine rhubarb and 2 cups of the water and cook over moderate heat for about 15 minutes. Strain in a sieve, rubbing pulp through. Measure out 2 cups and add the orange concentrate. Add remaining ⅓ cup water. Set over high heat and bring to a full, rolling boil, stirring frequently. Stir in pectin and bring again to full boil, stirring constantly. Boil 1 minute. Remove from heat and skim. Seal in hot sterilized glasses. About 20 pints.

Pineapple-Rhubarb Jam

5 cups (½-inch pieces) rhubarb
4 cups sugar
1 (8 ounce) can crushed pineapple
1 (6 ounce) package strawberry-flavored gelatin

Put rhubarb, sugar, and pineapple with juice in a large kettle or 5-quart Dutch oven. Cook over medium heat for a few minutes, stirring often, to combine ingredients. When sugar has dissolved, turn up heat and cook at a slow boil for 30 minutes. Remove from heat and stir in gelatin until dissolved. Put into 6 half-pint jars, seal and refrigerate or freeze.

Rhubarb-Raspberry Jam

5 cups finely cut rhubarb
1 cup water
5 cups sugar
½ teaspoon salt (optional)
1 can raspberry pie filling
1 3-oz. package raspberry gelatin
1 3-oz. package lemon gelatin

Cook rhubarb and water until tender. Add sugar. Cook and stir 3-4 minutes longer. Add pie filling and cook 6-8 minutes longer on low heat. Remove from heat and add gelatin. Pour into jars and seal. Store in refrigerator or freezer.

Fresh Strawberry-Rhubarb Jelly

1½ pounds unpeeled raw rhubarb
1 quart ripe strawberries, hulled
1 (16 ounce) package powdered fruit pectin
5 cups sugar

Put rhubarb through food grinder, saving all juice. Mash strawberries and add to rhubarb. Place in a jelly bag to drain, pressing out juice. Measure: there should be 3½ cups. Add pectin and set over high heat. Bring to a full, rolling boil, stirring constantly. Add sugar and bring again to full boil, stirring constantly. Boil hard 1 minute. Remove from heat and skim. Seal in hot sterilized glasses. About 8 medium glasses.

Sweet 'n Low Strawberry-Rhubarb Jam

2 quarts fresh strawberries, hulled
  and halved
2 pounds rhubarb, trimmed and
  cut into 1" pieces
½ cup granulated sugar
3 tablespoons granulated sugar
  substitute
2 tablespoons bottled lemon juice
1 cup water
3 envelopes unflavored gelatin

In a large pot, combine all ingredients except water and gelatin. Over medium-high heat, bring to a boil. Reduce heat to medium and cook, stirring, 30-40 minutes or until very soft and slightly thickened. Meanwhile, in a small saucepan, sprinkle gelatin over water to soften. Over low heat, cook just until gelatin dissolves. Stir into fruit mixture. Funnel into hot sterilized jars, allowing ¼ head space (or follow jar manufacturer's instructions if different). Wipe rims with clean cloth dipped in hot water. Close according to manufacturer's instructions and process in boiling water bath 10 minutes. Refrigerate at least several hours prior to serving. Makes nine 8-ounce jars. 5 calories per tablespoon.

Rhubarb Conserve

2 pounds rhubarb
2½ pounds sugar
½ pound raisins
   Grated rind and juice of 1 orange
   Grated rind and juice of ½
   lemon

Wash the rhubarb and cut it into 1-inch pieces. Put in a pot with the remaining ingredients, mix well, cover, and let stand for 30 minutes. Bring to a boil, then simmer for 45 minutes, stirring frequently. Pour into hot, sterilized jars and seal. Makes six 6-ounce jars.

Pineapple-Rhubarb Conserve

8 cups chopped pineapple
8 cups peeled finely cut rhubarb
  Juice of one orange
  Juice of one lemon
6 pounds sugar

Combine all ingredients in a large pot. Stir over medium heat until sugar is dissolved. Bring to a boil; boil over medium heat for 45 minutes, or until the mixture is thick and sheets from spoon. Pour into hot sterilized jars; seal at once.

Strawberry-Rhubarb Conserve

3 cups sugar
2 cups diced rhubarb
1 cup seedless raisins
  Grated rind of two oranges
4 cups hulled strawberries
¼ cup chopped walnuts
  Hot sterilized jars
  Hot paraffin wax

Combine sugar, rhubarb and raisins with the rind of the oranges and let the mixture set for 12 hours. Next day, put in an enameled kettle with the strawberries and let the mixture cook slowly, stirring frequently until thick. Stir in the walnuts and out the conserve into hot sterile jars and seal with ¼ inch of hot paraffin wax. Store in a cool place.

Rhubarb-Raisin Marmalade

15 cups rhubarb, cut
7 oranges (4 peeled and 3
   unpeeled)
6 cups sugar
2 pounds raisins

Grind oranges. Mix all ingredients together. Put into a crock or enameled pan. Let stand overnight in refrigerator. Cook 20 minutes. Freeze or seal in jars. Makes eight pints.

Rhubarb-Carrot Marmalade

6 cups diced, peeled, raw rhubarb
3 cups ground, peeled, raw carrots
2 medium oranges, unpeeled
4½ cups sugar

Combine rhubarb and carrots. Put oranges through the food grinder, using a medium knife. Discard seeds but reserve all juice. Add to rhubarb mixture, then add the sugar. Let stand overnight. Stir over low heat until boiling; reduce heat and simmer until thickened — about 2 hours. Stir frequently. Seal in hot sterilized jars. Makes about 5 pints.

Rhubarb Fig Marmalade

1 pound rhubarb, cut fine
1 pound sugar
2 pound dried figs, cut small
   Juice of ½ lemon

Combine the ingredients in a large pot, cover, and let stand 24 hours. Cook rapidly, stirring frequently, until the jellying point is reached. Pour into hot, sterilized jars and seal. Makes six 6-ounce jars.

Rhubarb-Zucchini Conserve

3½ cups rhubarb, coarsely chopped
1 cup apple juice
½ cup honey
½ teaspoon minced fresh ginger
1 cup zucchini, coarsely chopped
¼ teaspoon grated fresh nutmeg
½ cup raisins
¼ cup walnuts, chopped

In a large saucepan, combine rhubarb, apple juice, honey and ginger. Cook rhubarb mixture on medium heat until thickened (about 15 minutes). Add zucchini and cook five minutes longer. Remove from stove. Add nutmeg, raisins and walnuts. Pour into hot sterilized half-pint jars. Seal and refrigerate. Can be used as a dip or condiment for poultry.

Rhubarb Marmalade

2 pounds rhubarb
2 lemons
5 cups sugar

Cut the rhubarb finely. Remove outer (yellow) half of the lemon rinds and slice finely. Mix rhubarb, lemon rind, and sugar, and let stand overnight. Next day squeeze lemons, strain juice, add juice to mixture, and cook until quite thick, stirring almost constantly to prevent burning. Turn into sterilized jars and seal.

Sauces & Soups

Basic Rhubarb Sauce

Microwave Rhubarb Sauce

Frozen Rhubarb Sauce

Oven-Easy Rhubarb Sauce

Honey-Rhubarb Syrup

Strawberry-Honey-Rhubarb Sauce

Baked Rhubarb

Springtime Rhubarb Refresher

Gelatin-Rhubarb Sauce

Honeyed Rhubarb and Pear Compote

Rhubarb Pineapple Sauce

Crock Pot Rhu-berry Sauce

Springtime Sundae

Strawberry-Honey-Rhubarb Sauce

Cranberry Consommé

Rhubarb Borscht

Rhubarb Soup

Creamy Rhubarb Soup

Basic Rhubarb Sauce

Wash rhubarb; peel if skin is tough. Cut in 1-inch pieces and measure 3 cups. Add 1 cup sugar and very small amount of hot water. Cover and cook slowly until tender, about 25 minutes. Makes 6 servings.

Microwave Rhubarb Sauce

    3 cups rhubarb, cut ½-inch thick
    ¼ cup water
    ¾-1 cup sugar

Combine rhubarb and water in microwave-safe 2-quart measure or casserole. Microwave on High, covered, 5 to 7 minutes, stirring once during cooking. Add sugar; stir well to dissolve. Cover and cool slightly or chill before serving. Makes 4 to 6 servings.

Frozen Rhubarb Sauce

    3 cups frozen rhubarb
    1 cup granulated sugar
    ¾ cup water

Combine rhubarb, sugar and water in saucepan. Cover and cook over low heat until rhubarb is defrosted and mixture comes to a boil, stirring once or twice. Makes about 3 cups.

Oven-Easy Rhubarb Sauce

    3 cups cut-up rhubarb
    ¾ cup granulated sugar
    2-3 tablespoons orange juice or
        water

Place rhubarb in small casserole. Cover with the sugar and toss to coat evenly.

Pour on juice or water to moisten sugar. Bake, uncovered, at 350° F. for 30 minutes or until just tender. Makes 3 cups.

Honey-Rhubarb Syrup

    1 pound rhubarb
    1 cup honey (or more to taste)
    ¼ cup water

Combine honey and water in large saucepan and bring to slow boil. Cut rhubarb into 1-inch pieces and add to pot. Cook slowly until rhubarb is tender, 25-30 minutes. Serves 3 or 4.

Strawberry-Honey-Rhubarb Sauce

    3 cups diced rhubarb
    1 cup boiling water
    1 cup sugar
    1 two-inch stick cinnamon

Cover the rhubarb with boiling water, let stand five minutes, then drain. Meanwhile make a syrup with the cup of boiling water, the sugar, and the cinnamon. Add the drained fruit and cook gently until the fruit is tender but not broken. *For Stewed Rhubarb and Strawberries, prepare the rhubarb as directed above.* When almost tender, add one-half pint of strawberries (washed, hulled and halved) and cook three minutes longer. Serves 4 to 6.

Baked Rhubarb

    2 lbs. rhubarb, cut into 1" pieces
    1 cup sugar
        Dash salt
        Dash cinnamon

Add the salt and cinnamon to the sugar

and mix with the rhubarb. Place in a large flat glass baking dish without adding any water. Bake in a 350° F. oven until tender, about ½ hour. There will be plenty of juice. Serve warm or cold with heavy cream. Serves 6.

Springtime Rhubarb Refresher

- 1 **pound rhubarb, cut into one-inch pieces (about 2½ cups)**
- ¼ **cup sugar**
- ¼ **cup strawberry jelly**
- ¼ **teaspoon ground cinnamon**
- ⅛ **teaspoon salt**

In a two-quart saucepan over medium heat, heat all ingredients to boiling. Reduce heat to low; cover and simmer 10 minutes or until rhubarb is tender, stirring the mixture occasionally. Spoon the rhubarb mixture into four dessert dishes. Serve the rhubarb warm or refrigerate to serve cold later.

Gelatin-Rhubarb Sauce

- 3-4 **cups chopped 1-inch rhubarb**
- ½-¾ **cup sugar**
- 2 **tablespoons water**
- 1 **package strawberry gelatin**

Combine rhubarb with sugar, water, and gelatin in medium saucepan. Bring to a boil. Reduce heat and cover, stirring occasionally as it simmers, for 10 minutes or until cooked up for sauce. Don't add sugar right away unless you want the rhubarb to stay firmer. Makes about 2 cups.

Honeyed Rhubarb and Pear Compote

- 1½ **lbs. rhubarb**
- ⅓ **cup liquid honey**
- ⅓ **cup sugar**
- 2½ **cups pear halves, no. 2 can**

Cut rhubarb into two-inch pieces. Place in shallow baking dish. Mix honey and sugar. Pour over rhubarb. Cover and bake in moderate oven (350°) for 40 minutes or until tender, stirring after 20 minutes. Chill. Arrange in serving dish. Drain pears and place on rhubarb. Serves 4-6.

Rhubarb Pineapple Sauce

- 3 **cups rhubarb, diced**
- 1 **8½ oz. can crushed pineapple**
- ½ **cup sugar**
- ½ **cup water**
- ¼ **cup red cinnamon candies**
- 2 **tablespoons cornstarch**
- 2 **tablespoons water**
- ¼ **teaspoon salt**
- 2 **tablespoons lemon juice**
- 2 **tablespoons butter or margarine**

Combine rhubarb, undrained pineapple, sugar, ½ cup water and cinnamon candies in a two-quart saucepan; mix well. Cook over medium heat until mixture comes to a boil, stirring occasionally. Combine cornstarch and two tablespoons water; mix to blend. Gradually stir into hot mixture. Reduce heat to low; simmer for five minutes. Remove from heat. Stir in salt, lemon juice and butter. Cool slightly before serving. Delicious served over ice cream or squares of cake. Or pour slightly-cooked sauce into jars or bottles; cover with lids. Store in refrigerator. Reheat stored sauce before using. Makes four cups.

Crock Pot Rhu-berry Sauce

4 cups thinly sliced rhubarb
1 (10 ounce) package frozen
  raspberries or strawberries,
  thawed
¾ cup granulated sugar
3 tablespoons Minute tapioca
1 teaspoon grated orange peel
¼ cup water
  Light cream or ice cream

Place the rhubarb and berries in the crock; stir to mix. Stir together the sugar, tapioca, orange peel and water; mix into fruit. Let stand 15 minutes. Cover cooker. Cook on low setting 3-4 hours or until rhubarb is just tender when pierced with a fork. Makes 6 to 8 servings.

Springtime Sundae

2 cups diced rhubarb
1 tablespoon water
1 cup sugar
1 pint strawberries, washed, hulled
  and sliced
1 quart vanilla ice cream
  Toasted almonds

Cook rhubarb and water in a covered saucepan until rhubarb is tender, about 5 minutes. Add sugar and sliced strawberries and simmer 3 minutes, or until strawberries are soft. Remove from heat and chill thoroughly. Serve a generous amount over the ice cream and the rest in a sauce dish. Top with toasted almonds. Makes 3 cups sauce.

Strawberry-Honey-Rhubarb Sauce

2½ cups rhubarb,
   cut ½-inch thick
1½ cups sliced fresh strawberries
⅓-½ cup honey
¼ cup water

Combine rhubarb, strawberries, ⅓ cup honey and water in microwave-safe 2-quart measure or casserole. Microwave on High, covered, 5 to 7 minutes, stirring once during cooking, or until fruit is tender. Cover and cool slightly or chill before serving. Makes 5 servings.

Cranberry Consommé

1 16-ounce package frozen
  unsweetened sliced rhubarb
¾ cup sugar
3 inch stick cinnamon
2 cups cranberry juice
½ cup burgundy
½ cup club soda
  Fresh mint (optional)

In a saucepan combine rhubarb, sugar, cinnamon, and two cups water. Bring to boiling; reduce heat and simmer five minutes or until rhubarb is tender. Remove cinnamon stick. Pour rhubarb into strainer and press out juice. Add cranberry juice and wine to rhubarb juice. (If desired, chill juice mixture until serving time. Just before serving, in a saucepan heat just until warm.) Add club soda; mix gently. Garnish with fresh mint, if desired. Serve warm.

Rhubarb Borscht

2 cups puréed rhubarb
½ cup water
6 tablespoons orange juice
2 eggs, beaten
5 tablespoons sour cream
3 tablespoons sugar (or to taste)

Mix all ingredients until well blended. Chill well. Serve with a dollop of yogurt.

Rhubarb Soup

2 pounds rhubarb, diced
3½ cups water
4 (2-inch) sticks cinnamon
¼ teaspoon salt
1½ tablespoons cornstarch
1 cup sugar
Red food coloring
1 slice lemon or lime

Combine rhubarb, water, cinnamon and salt in saucepan. Cover and cook 20 minutes or until rhubarb is very tender. Remove cinnamon sticks. Put half of mixture through sieve, or purée in blender. Return to pan. Blend cornstarch with sugar and stir into mixture. Cook, stirring constantly, until juice begins to thicken. Remove from heat and tint with food color as desired. Add lemon slice. Cool, then chill, covered. Serve with cake or cookies, or topped with sour cream or whipped cream. Make about 6 cups.

Creamy Rhubarb Soup

3 pounds rhubarb, cut into 1-inch pieces
½ cup sugar
¼ teaspoon salt
½ cup water
1 pound frozen strawberries
¾ teaspoon cinnamon
¼ cup water
¼ cup cornstarch
2 cups milk
3 cups sour cream

Cook rhubarb, sugar, salt and ½-cup water until sauce-like in consistency. Combine water and cornstarch; combine strawberries and cinnamon. Add two mixtures together in saucepan and bring to a boil. Simmer gently 1 minute. Combine with rhubarb sauce. Chill. When ready to serve, add milk and sour cream and mix well. Makes 10 servings.

Puddings & Soufflés

Rhubarb Tapioca
Rhubarb Cream Pudding
Rhubarb Whip
Rhubarb Sponge Custard
Dutch Rhubarb Cream
Fluffy Rhubarb Fool
Rhubarb Tansy
Banana-Rhubarb Pudding
Rhubarb Bread Pudding
Rhubarb Date Pudding
Rhubarb Mousse
Rhubarb Scallop with Meringue
Strawberry-Rhubarb Mousse
Maple-Rhubarb Tofu Mousse
Ginger Soufflé with Rhubarb-Ginger Sauce
Rhubarb-Ginger Sauce
Cold Rhubarb Soufflé

Rhubarb Tapioca

3 tablespoons quick-cooking tapioca
1 cup sugar
¼ teaspoon salt
1 cup water
3 cups rhubarb, cut into ½-inch pieces.
¼ cup heavy cream, whipped
½ teaspoon grated orange rind

Combine tapioca, sugar, salt, water and rhubarb in saucepan and bring to a boil, stirring constantly. Cook over low heat for two minutes. Chill. Serve with whipped cream flavored with orange rind. If desired, instead of whipped cream, use chilled soft custard or unwhipped heavy cream, flavored with nutmeg. Serves four.

Rhubarb Cream Pudding

2 tablespoons cornstarch
½ cup sugar
⅛ teaspoon salt
2 cups milk
3 eggs, beaten

Rhubarb sauce:
1 pound of rhubarb
1 tablespoon water
½ cup sugar
1 tablespoon cornstarch

Mix cornstarch, sugar, and salt in top of a double boiler. Add cold milk and stir until smooth. Cook over direct heat, stirring constantly until mixture boils and thickens. Slowly stir part of the hot mixture into the beaten eggs. Return to hot mixture in top of double boiler and continue cooking over boiling water for two minutes, stirring constantly to keep smooth. Remove from heat and cool. Prepare rhubarb sauce. Wash rhubarb, discard stem ends and leaves. Slice stalks in one-inch lengths. Add water, cover, and cook slowly until rhubarb is soft. Mix sugar and cornstarch thoroughly, and stir into the rhubarb; stir constantly over low heat until mixture boils and thickens. Chill, and when ready to serve, stir rhubarb sauce into chilled cream pudding. Serves 5.

Rhubarb Whip

4 cups diced rhubarb
¼ cup water
Sugar (sweeten to taste)
2 cups heavy cream, stiffly beaten
Lady fingers or sponge cake

Simmer the rhubarb gently with the water until soft, add the sugar, and when this is dissolved, press through a fine sieve and chill. Just before serving, fold in the whipped cream and serve in sherbet glasses with split lady fingers or strips of sponge cake around the edge. Serves 6.

Rhubarb Sponge Custard

1 cup diced fresh rhubarb
¼ cup water
½ cup sugar
¼ cup all-purpose flour
¼ teaspoon salt
¼ cup sugar
2 tablespoons butter
2 eggs, separated
1 tablespoon lemon juice

Put rhubarb in saucepan, add water and sugar. Cover and cook until tender.

(Makes about one cup thick sauce.) Sift flour, measure and resift with remaining dry ingredients. Add to the creamed butter and mix well. Add the beaten egg yolks and lemon juice and beat until light and fluffy. Add cooked rhubarb and mix well. Beat the egg whites until stiff. Fold lightly but thoroughly into the rhubarb mixture. Pour into custard cups and set in a pan of hot water that comes almost to the top of the cups. Bake in a moderately slow oven (325°) for 30 minutes. Sauce will separate and remain at bottom with a fluffy cake-like topping. Serve warm or chilled.

Dutch Rhubarb Cream

     1 **pound rhubarb, cut in 1" pieces**
  1½ **cups water**
     1 **cup sugar**
     1 **teaspoon grated lemon rind**
     2 **envelopes unflavored gelatin**
  ½ **cup heavy cream**

Combine the rhubarb, one cup of the water, sugar and lemon rind in a medium-sized saucepan. Cover and bring to a boil; lower heat and simmer until tender. Sprinkle gelatin over the ½ cup water in a one cup measure; let stand five minutes to soften. Stir into hot mixture. Cook five more minutes, mashing rhubarb. Pour into bowl; chill until mixture will hold its shape softly when spooned. Beat cream in a small bowl with electric mixer until stiff. Fold into rhubarb until no white streaks remain. Spoon into five-cup dish or individual sherbets. Chill four hours or until soft-set. Or chill longer and remove from refrigerator 30-60 minutes before serving.

Fluffy Rhubarb Fool

     3 **cups rhubarb in 1-inch pieces**
     1 **cup sugar, divided**
     2 **tablespoons cornstarch**
  1½ **cups milk**
     4 **eggs, separated**
     1 **teaspoon vanilla**
  ¼ **teaspoon cream of tartar**

In a covered saucepan over low heat, cook rhubarb (without water) until very tender, about 25 to 30 minutes. Stir in ½ cup sugar. Cook, uncovered, 5 minutes longer. Chill. In medium saucepan, combine ¼ cup of the sugar and the cornstarch. Gradually stir in milk. Cook over medium heat, stirring constantly, until mixture is thick. Remove from heat. Beat egg yolks and add vanilla. Beat hot mixture into the egg yolks. In a large mixing bowl, beat egg whites and cream of tartar until foamy. Add remaining ¼ cup of sugar, a tablespoon at a time, beating constantly until sugar is dissolved and whites are glossy and stand in soft peaks. Rub a bit of meringue between thumb and forefinger to feel whether sugar is dissolved. Gently but thoroughly fold yolk mixture into whites. Add rhubarb mixture, stirring just slightly to swirl colors. Pour into serving dishes. Makes 6 to 8 servings.

Rhubarb Tansy

- 1 pound rhubarb, cut in 1" pieces
- 4 ounces butter
- 2 ounces sugar
- 2 egg yolks, beaten
- ¼ pint double cream
- 2 tablespoons lemon juice
  Sugar, to taste

Simmer rhubarb gently in the butter until cooked. Add the beaten egg yolks and the lightly-whipped cream; sweeten to taste. Boil the mixture gently until it is barely firm. Turn out immediately into a serving dish. Sprinkle with sugar and lemon juice. Serve hot or cold.

Banana-Rhubarb Pudding

- 3 cups rhubarb, cut ½-inch thick
- 1 medium-sized ripe banana, sliced
- ¾ cup water
- ¼ cup sugar
- 3 tablespoon quick tapioca

Combine rhubarb, banana, water, sugar and tapioca in microwave-safe 2-quart measure or casserole. Microwave on High, covered, 7 to 10 minutes or until fruit is tender, stirring once during cooking. Cover and let stand until tapioca turns transparent. Serve at room temperature or cold. Makes 5 servings.

Rhubarb Bread Pudding

- 2½ cups diced rhubarb
- 4 cups milk
- 3 eggs
- 1½ cups sugar
- 1 teaspoon cinnamon
- ½ teaspoon coriander
- 1 pound loaf of french bread
- 4 tablespoons butter, melted

Tear bread into fairly small pieces and let stand an hour to partially dry. Mix well the milk, eggs, sugar, cinnamon, coriander and add rhubarb. Spread bread in well-buttered three-quart baking dish. Add rhubarb mixture and butter. Stir to mix. Bake at 325° for about an hour.

Rhubarb Date Pudding

- 2 cups diced rhubarb
- 1 cup chopped, pitted dates
- ¼ cup water
- ½ cup sugar
- 1 cup soft bread crumbs
- 1 teaspoon butter
- 6 quartered marshmallows

Cook together the rhubarb, dates, and water for about ten minutes. Add the sugar, crumbs, and butter, and turn into a buttered baking dish. Top with the marshmallows and bake in a moderate oven (350°) for about 20 minutes.

Rhubarb Mousse

- 1¼ cup rhubarb, washed, trimmed and cut into ½ pieces
- 8 tablespoons sugar
- 2 egg yolks, at room temperature
- 1 cup whipping cream
  Strawberries for garnish

Combine rhubarb and three tablespoons of the sugar in a heavy saucepan. Cook over very low heat, stirring occasionally until rhubarb dissolves into "strings." Remove from heat. Combine yolks and remaining five tablespoons of sugar in a double boiler. Whisk over simmering water until mixture thickens and begins to coat whisk. Set pan over a bowl of ice water. Add 2 tbsp. rhubarb mixture and beat until cooled. Whip cream until soft peaks form. Fold yolk mixture into whipped cream. Gently fold in remaining rhubarb mixture. Divide among serving glasses. Refrigerate. Makes 4 servings.

Rhubarb Scallop with Meringue

- ½ pound rhubarb, cut in 1" pieces
- 1 cup granulated sugar
  Grated rind of 1 orange
- ¼ teaspoon salt
- 1 small sponge cake
- 2 egg whites
- 2 tablespoons powdered sugar

Combine rhubarb, sugar, orange rind, and salt. Line bottom of baking dish with 3-4 slices of pound cake; cover with rhubarb. Continue to make alternate layers of cake and fruit. Cover and bake in moderate oven (350°) for 30 minutes. Beat egg whites until stiff, add sugar slowly, beating until blended. Pile on baked pudding and bake 15 minutes longer, or until meringue is slightly brown. Serves 6.

Strawberry-Rhubarb Mousse

- 2 cups unsweetened rhubarb purée
- ½ cup sugar
- 3 egg whites
- 3 tablespoons water
- 1 tablespoon unflavored gelatin
- ½ cup finely chopped strawberries
- 1 cup whipping cream

Soften gelatin in water. Combine purée and sugar. Heat until sugar is completely dissolved. Add gelatin and stir until gelatin is dissolved. Cool the mixture. Beat the egg whites to form stiff peaks. In another bowl, beat the cream until stiff. Stir ⅓ of the whites into the rhubarb mixture, then fold in the rest. Fold in the whipped cream and strawberries until only a few streaks remain. Refrigerate until set.

Maple-Rhubarb Tofu Mousse

- 2 tablespoons agar-agar flakes
- 2 cups fruit juice (apple, berry, etc.)
- 1 cake of tofu
- ¾ cup sunflower seeds, pulverized
- ⅓ cup maple syrup
- ¾ cup oil
- 1 teaspoon lemon juice
- ¾ teaspoon salt
- 1 teaspoon vanilla extract
- 2 cups rhubarb
- ¼ cup maple syrup
- ¼ cup apple-strawberry juice

Mix the agar-agar flakes with 2 cups of juice. Simmer. Add one cake of tofu and cook gently for 10 minutes. Lift the tofu out and set it and the liquid aside. Combine the seeds with ⅓ cup maple syrup, adding the oil alternately with pieces of drained tofu. When well-mixed, scrape down and add the lemon juice, salt and vanilla. Turn into food processor and add the agar-agar mixture and juice. Lightly oil a ring mold or 7 individual custard molds. Fill the molds and refrigerate. Dice the rhubarb and put into a pot with ¼ cup maple syrup and ½ cup fruit juice. Let stand ½ hour and then simmer until rhubarb is done. Cool and refrigerate. Use a knife to loosen edges of mousse. Turn out ring mold onto a serving platter and top with the rhubarb sauce.

Ginger Soufflé with Rhubarb-Ginger Sauce

Rhubarb-Ginger Soufflé Sauce
(recipe follows)

6 tablespoons (¾ stick) unsalted butter
6 teaspoons unbleached all-purpose flour
1 cup milk
½ cup heavy or whipping cream
5 egg yolks
½ cup sugar
½ cup finely chopped crystallized ginger
1 teaspoon orange-flower water
7 egg whites at room temperature
Pinch cream of tartar

Make the rhubarb sauce. Melt the butter in a heavy small saucepan over medium heat until foamy. Stir in the flour and cook 1 minute. Gradually stir in the milk and cream. Cook, stirring constantly, until thick and smooth. Remove from heat. Add the egg yolks, one at a time, whisking well after each addition. Stir in the sugar, then the ginger and orange-flower water. Preheat oven to 450°. Butter a 6-cup souffle dish and coat with granulated sugar. Beat the egg whites with the cream of tartar until stiff but not dry. Gently fold into the souffle base. Pour the batter into the prepared dish. Bake until puffed and golden, about 30 minutes. Serve immediately with rhubarb sauce spooned around each serving. Serves 6.

Rhubarb-Ginger Sauce

3 cups chopped rhubarb
⅓ cup sugar
⅓ cup orange liqueur
⅓ cup (or as needed) water
2 tablespoons finely chopped crystallized ginger

Combine the rhubarb, sugar, orange liqueur, and ⅓ cup water in a heavy large saucepan. Heat to boiling. Reduce heat and simmer uncovered, stirring occasionally, until as thick as applesauce, about 30 minutes. Stir in the ginger and simmer another 15 minutes, adding more water if the sauce is too thick. Remove from heat and let cool to room temperature. Makes about 1½ to 2 cups.

Cold Rhubarb Soufflé

4 cups chopped rhubarb
2¼ cups sugar
1 envelope gelatin
1 cup heavy cream
4 egg whites
1 teaspoon vanilla
12 fresh strawberries
Additional whipped cream

Cook the rhubarb with ¼ cup water and 1¾ cups of the sugar in a heavy saucepan for about 10 minutes until soft. Strain and cook down the juice to ½ cup. Purée the rhubarb in a food processor or blender, or put through a vegetable mill. Soften the gelatin in two tablespoons cold water, then add to the hot rhubarb juice and stir until completely dissolved. Add the purée. Beat the cream until stiff. Beat the egg whites until they begin to stiffen, then add the remaining ½ cup sugar and the vanilla, continuing to beat until stiff peaks form. Fold first the egg-white mixture into the rhubarb, then the whipped cream. Make a collar of waxed paper and fit it around a 1½ quart soufflé mold. Chill at least six hours. Remove the collar and decorate the top with fresh strawberries and rosettes of whipped cream piped from a pastry tube.

Muffins & Breads

Rhubarb Muffins

Rhubarb Maple Muffins

Rhubarb Mini-Muffins

North Dakota Rhubarb Muffins

Crunchy Rhubarb Muffins

Refrigerator Bran Muffins with Rhubarb

Rhubarb Bran Muffins

Strawberry-Rhubarb Muffins

Rhubarb Sticky Buns

Rhubarb Nut Bread

Honey Rhubarb Bread

Rhubarb Nature Bread

Banana-Rhubarb Bread

Orange Rhubarb Bread

Rhubarb-Orange Bread

Rhubarb Muffins

¾ pounds rhubarb, chopped
2½ cups unsifted all-purpose flour
½ cup chopped walnuts or pecans
1 teaspoon baking powder
1 teaspoon baking soda
½ teaspoon salt
¼ teaspoon nutmeg
1 cup firmly packed light brown sugar
1¼ cups buttermilk
½ cup vegetable oil
1 large egg
2 teaspoons vanilla

Heat oven to 375°. Grease 18 2½" muffin pan cups. In large bowl, mix flour, nuts, baking powder, soda, salt and nutmeg. In small bowl beat brown sugar, milk, oil, egg and vanilla. Stir into flour mixture just until moistened (batter will be lumpy). Fold rhubarb into batter. Spoon batter into cups. Bake 18-20 minutes. Cool in pan five minutes. Serve immediately.

Rhubarb Maple Muffins

1½ cups diced rhubarb
⅓ cup sugar
2½ cups flour
1 tablespoon baking powder
1 teaspoon salt
½ teaspoon cinnamon
½ teaspoon allspice
¾ stick (6 tablespoons) butter
½ cup maple syrup
1 egg
½ cup milk

Combine rhubarb and sugar and let stand 1 hour. Combine flour, salt, baking powder, allspice. Cream butter, syrup, egg; blend in milk. Mix in rhubarb and add dry ingredients, mixing only until combined. Bake at 400°. Makes 12 muffins.

Rhubarb Mini-Muffins

¼ cup packed brown sugar
½ teaspoon cinnamon
¼ cup finely chopped pecans
1½ cups flour
½ teaspoon baking soda
½ teaspoon salt
½ teaspoon cinnamon
½ teaspoon nutmeg
¾ cup packed brown sugar
½ cup buttermilk
⅓ cup vegetable oil
1 egg, slightly beaten
1 teaspoon vanilla
1 cup finely chopped rhubarb
½ cup finely chopped pecans
36 pastel-colored nut cups

Preheat oven to 325°. In small bowl combine ¼ cup brown sugar, ½ teaspoon cinnamon and ¼ cup pecans; set aside. In medium bowl, combine flour, soda, salt, ½ teaspoon cinnamon and nutmeg. Beat ¾ cup brown sugar, buttermilk, oil, egg and vanilla together; stir into dry ingredients until almost blended. Gently fold in rhubarb and ½ cup pecans. Fill nut cups ⅔ full; sprinkle with about ½ teaspoon reserved pecan mixture. Arrange filled cups on baking sheet so sides do not touch. Bake until muffins test done with wooden pick, 18-19 minutes. Remove to wire rack to cool. Make 36 mini-muffins.

North Dakota Rhubarb Muffins

1¼ cups brown sugar
½ cup oil
1 egg
2 teaspoons vanilla
½ cup nuts
1 cup milk
2 cups finely diced rhubarb
2½ cups flour
1 teaspoon baking powder
1 teaspoon soda
1 teaspoon salt
⅓ cup white sugar
½ teaspoon cinnamon
1 teaspoon margarine

Beat brown sugar and oil. Add egg, vanilla, milk, rhubarb and nuts and blend well. Add flour, baking powder, soda and salt and mix until just blended. Fill greased muffin cups ⅔ full. Mix white sugar, cinnamon and margarine and sprinkle over batter in muffin cups, pressing in slightly. Bake at 375° for 20 minutes.

Crunchy Rhubarb Muffins

¾ cup brown sugar
½ cup oil
1 egg
½ cup buttermilk or ½ tablespoon vinegar with enough milk to make ½ cup
½ teaspoon salt
1 teaspoon vanilla
1½ cups flour
½ teaspoon soda
½ cup chopped nuts
1 cup finely chopped rhubarb (uniform size)

Topping:
½ cup brown sugar
½ cup nuts
¼ teaspoon cinnamon

Combine sugar, oil, egg, milk and vanilla. Mix well; add dry ingredients; add rhubarb and nuts. Put into greased muffin tins (not paper cups). Sprinkle topping on top of batter. Bake at 325° for 30 minutes. Makes 12 muffins.

Refrigerator Bran Muffins with Rhubarb

2 cups 100% Bran
2 cups boiling water
3 cups plus 2 tablespoons sugar
1 cup plus 2 tbsp. shortening
4 eggs
4 cups All-Bran
5 cups flour
1 tablespoon salt
5 teaspoons soda
½ teaspoon baking powder
1 quart buttermilk
Rhubarb, cooked sweetened, allow ½ tablespoon per muffin

Pour water over bran and let set. Cream together sugar and shortening, then add eggs to creamed mixture. Add All-Bran to creamed mixture, mixing well. Sift together flour, salt, soda and baking powder and add alternating with buttermilk. Mix well. Add 100% Bran and water mixture, folding in carefully. Add ½ tablespoon cooked sweetened rhubarb to each cup in a muffin tin, mixing slightly in the tin. Bake at 400° for 15 minutes. Batter will keep in the refrigerator for up to six weeks.

Rhubarb Bran Muffins

- 1 cup sour milk (add ½ tablespoon lemon juice)
- 1 cup All-Bran cereal
- 1 egg
- ½ cup oil
- 1½ cup flour
- ¾ cup sugar
- 1 teaspoon baking soda
- ½ cup rhubarb, diced
- 1 teaspoon cinnamon, mixed with
- 2 tablespoons sugar for topping

If needed, sour the milk and let it curdle. Place cereal and sour milk in mixing bowl. Let stand a few minutes to soften cereal. Add egg and oil; mix well. Add flour that has been combined with the sugar and soda. Stir only enough to moisten. Fold in rhubarb. Fill greased muffin cups ½-⅔ full. Sprinkle cinnamon-sugar mixture on top. Bake in 375° oven for 20-25 minutes. Let cool a few minutes before removing from pan. Serve warm or cold.

Strawberry-Rhubarb Muffins

- 1¾ cup flour
- ½ cup sugar
- 2½ teaspoons
- ¾ teaspoon salt
- 1 egg, lightly beaten
- ¾ cup milk
- ⅓ cup vegetable oil
- ¾ cup rhubarb, minced
- ½ cup strawberries, sliced
- 6 small strawberries, cut in half sugar

Combine flour, sugar, baking powder and salt in a large bowl. Combine egg, milk and oil in a small bowl; then stir into flour mixture with a fork until just moistened. Fold rhubarb and sliced strawberries into batter. Fill well-greased or paper-lined muffin tin cups ⅔ full of batter. Press strawberry half gently onto top of each muffin. Sprinkle tops generously with sugar. Bake in 400° oven for 20-25 minutes until golden.

Rhubarb Sticky Buns

- 1 package hot roll mix
- ¾ cup very warm water (105-115°)
- 2 tablespoons sugar
- 1 egg
- ¼ cup firmly packed brown sugar
- 1 cup frozen rhubarb
- ¼ cup margarine
- ¼ cup corn syrup
- 2 tablespoons sugar
- 2 teaspoons cinnamon

In a large bowl, dissolve yeast from hot roll mix in water. Stir in sugar and egg. Add flour mixture; blend well. Cover; let rise in warm place until light and doubled in size, 45-60 minutes. In medium saucepan, combine brown sugar, rhubarb, butter and corn syrup. Boil three minutes, stirring occasionally. Spoon into ungreased 13"x9" pan or 12 muffin cups. On well-floured surface, toss dough until no longer sticky. Press or roll out to 12"x6" rectangle. Combine sugar and cinnamon; sprinkle over dough. Starting with longer side, roll up tightly; cut into 12 slices. Place in prepared pan. Cover; let rise in a warm place until light and doubled in size (30-45 minutes). Preheat oven to 375°. Bake 20-25 minutes until golden brown. Cool five minutes; remove from pan. Makes 12 rolls.

Rhubarb Nut Bread

- 1½ cups brown sugar
- ⅔ cup salad oil
- 1 egg
- 1 cup buttermilk or 1 cup sour milk*
- 1 teaspoon salt
- 1 teaspoon soda
- 1 teaspoon vanilla
  Mix, then add to above mixture:
- 2½ cups flour
- 1½ cups diced rhubarb (fresh/frozen)
- ½ cup chopped nuts
  Topping:
- ½ cup sugar
- 1 tablespoon melted butter
- 1 tablespoon cinnamon

Pour batter into two greased loaf pans. Sprinkle topping over batter. Bake at 350° for 45 minutes. Don't overbake. Freezes very well. *To make sour milk, add 1 tsp. vinegar to 1 cup milk.

Honey Rhubarb Bread

- 3 eggs, beaten
- 1 cup oil
- 1½ cups honey
- 1 teaspoon vanilla
- 2 cups rhubarb, cooked and sweetened with ¼ cup honey
- 3 cups whole wheat flour
- 1 teaspoon salt
- 2 teaspoons baking soda
- ¼ teaspoon cinnamon
- ½ teaspoon nutmeg

Stir together first 5 ingredients; slowly add dry ingredients. Divide into 2 greased and floured loaf pans. Bake at 350° for 50 minutes to 1 hour. Cover with foil last 20 minutes.

Rhubarb Nature Bread

Combine:
- 3 eggs, well beaten
- 1 cup salad oil
- 2 cups brown sugar
- 2 teaspoons vanilla
  Add:
- 2½ cups diced rhubarb
- ¾ cup chopped nuts
  Combine separately:
- 1½ cups white flour
- 1½ cups whole wheat flour
- 2 teaspoons soda
- 2 teaspoons cinnamon
- 1 teaspoon salt
- ½ teaspoon baking powder
- ½ teaspoon nutmeg
- ½ teaspoon allspice

Combine both mixtures and stir gently. Pour into two 5"x9" pans (greased). Bake for one hour at 350°. Cool for 10 minutes before turning bread out on racks.

Banana-Rhubarb Bread

1 orange
  boiling water
2 tablespoons margarine
1 egg
1 cup sugar
1½ cups rhubarb, chopped
1 teaspoon cardamon, ground
1 banana, mashed
1 cup flour
¾ cup whole wheat flour
½ teaspoon salt
½ teaspoon baking soda
¼ cup pecans

Squeeze the juice from one orange and add boiling water to make ¾ cup liquid. Stir in margarine until melted. In a large bowl beat sugar and egg. Add orange juice mixture, rhubarb, banana and cardamon. Sift dry ingredients, add to liquid. Stir in chopped nuts. Spoon into buttered bread pan. Bake one hour and fifteen minutes at 325°. Makes one loaf.

Orange Rhubarb Bread

1 tablespoon grated orange rind
½ cup orange juice
3 tablespoons frozen orange concentrate
3 tablespoons butter, melted
2 eggs, beaten
½ teaspoon salt
2½ cups sifted flour
½ teaspoon salt
1 teaspoon cinnamon
2 cups diced rhubarb
⅔ cups sugar

Beat orange rind, juice, concentrate, butter and eggs together. Sift dry ingredients and add. Stir in rhubarb and turn into greased loaf pan. Bake at 350° for one hour.

Rhubarb-Orange Bread

½ cup shortening
⅔ cup sugar
2 eggs
2 cups all-purpose flour
1 teaspoon baking powder
½ teaspoon soda
1 teaspoon salt
½ cup orange juice
1 cup finely cut-up rhubarb
½ cup chopped nuts
1 tablespoon grated orange peel

Heat oven to 350°. Grease bottom only of a 9x5x3-inch pan. In mixer bowl, cream shortening and sugar. Add eggs, one at a time. Combine flour, baking powder, soda and salt. Stir in alternately with orange juice. Stir in rhubarb, nuts and orange peel. Pour batter into pan; let stand 20 minutes. Bake 60 minutes or until wooden pick inserted in center comes out clean. Flavor improves after 1 day. Makes 1 loaf.

Cobblers, Bettys, Dumplings & Roll-ups

Rhubarb Cobbler

- 1 cup sugar
- ⅓ cup pancake mix
- 1 teaspoon grated lemon peel
- 4 cups rhubarb, cut in ½" pieces

Topping:
- ⅔ cup sugar
- ¾ cup pancake mix
- 1 egg, beaten
- ¼ cup melted butter

Preheat oven to 375°. Mix sugar, pancake mix, and lemon peel, add rhubarb and toss lightly. Place in nine-inch square pan. For topping, combine pancake mix and sugar. Stir in egg until mixture resembles coarse crumbs. Sprinkle evenly over rhubarb base. Drizzle with melted butter. Bake in 375° oven for 35 to 40 minutes. Serve warm with cream or ice cream.

Rhubarb Cobbler with Oat Dumplings

- ¾ cup sugar
- 2 tablespoons cornstarch
- 1 cup water
- ½ cup orange juice
- 1 pound rhubarb, cut into 1" pieces
  Dumplings:
- ½ cup all-purpose flour
- ¼ cup whole wheat flour
- ¼ cup quick cooking rolled oats
- 1½ teaspoons baking powder
- ½ cup skim milk
- 2 tablespoons cooking oil
  Topping:
- 1 tablespoon sugar
- ¼ teaspoon ground cinnamon

In a small saucepan stir together the sugar and cornstarch. Stir in the water and orange juice. Cook and stir over medium-high heat till the mixture is thickened and bubbly. Add the fresh or frozen rhubarb; cook and stir until the mixture returns to boiling. Remove from the heat. Cover the mixture to keep warm. For dumplings, in a medium mixing bowl stir together the all-purpose flour, whole wheat flour, rolled oats, and baking powder. In a 1 cup glass measure stir together the skim milk and cooking oil. Add the milk mixture to the flour mixture; stir just till flour mixture is moistened. Do not overmix. Transfer the warm rhubarb mixture to a 2-quart casserole. Immediately spoon the dumpling batter into 8 mounds on top of the warm rhubarb mixture. In a small dish stir together the remaining 1 tablespoon sugar and cinnamon; sprinkle over the top of the dumplings. Bake, uncovered, in a 425° oven about 20 minutes or until a toothpick inserted in the center of a dumpling comes out clean. Serve warm. Makes 8 dessert servings.

Rhubarb-Blueberry Cobbler

- 2 cups sliced rhubarb
- 1 cup frozen or fresh blueberries
- ⅔ cup granulated sugar
- ½ cup water
- 1 tablespoon cornstarch
- ¼ teaspoon cinnamon
- 1¼ cups buttermilk biscuit mix
- 1 tablespoon sugar
- ¼ cup low-fat milk
- 1 tablespoon margarine, melted
  Sugar and cinnamon (optional)
  Vanilla ice milk (optional)

Stir together rhubarb, blueberries, ⅔ cup sugar, water, cornstarch and cinnamon in a microwave-safe 2-quart round casserole. Cover with casserole lid or plastic wrap. Microwave on High for 8 to 9 minutes, stirring twice toward end of cooking, or until sauce is bubbly and thickened. While fruit cooks, make cobbler topping by combining biscuit mix and 1 tablespoon sugar. Add milk and melted margarine; stir with fork until well mixed. Drop by small spoonfuls around edges of hot fruit, leaving center open. Sprinkle with sugar cinnamon mixture, if desired. Cover with lid or waxed paper. Microwave on High 3½ to 4½ minutes or until topping no longer looks doughy on top (and when pulled apart with fork, inside parts are cooked). Serve warm in individual dishes, spooning fruit over cobbler topping. Top with ice cream, if desired. Makes 6 to 8 servings.

Easy Rhubarb Betty

    2 pounds trimmed rhubarb stalks, in 2-inch pieces
 1½ cups sugar
 2½ cups homemade dry bread crumbs
    6 tablespoons butter, melted

Preheat the oven to 350°. Place the rhubarb in a saucepan with the sugar and ¼ cup water, cover, and cook until tender, about 7 minutes. Set aside ⅓ cup of the rhubarb syrup. Put the bread crumbs in a bowl with the melted butter and toss. Place half the crumbs in the bottom of a baking dish. Arrange the rhubarb on top of the crumbs and pour the syrup over it. Top with the remaining crumbs. Bake about 40 minutes, covering with foil if the crumbs begin to get too brown. Serve with whipped cream.

Strawberry-Rhubarb Cobbler

 ¾ cup granulated sugar
    2 tablespoon cornstarch
 ⅛ teaspoon salt
    1 (10 ounce) package sliced strawberries, thawed
    3 cup sliced fresh or thawed frozen rhubarb
    1 tablespoon lemon juice
    1 tablespoon margarine or butter
    1 can (10 biscuits) refrigerated buttermilk or country style biscuits
    1 tablespoon sugar mixed with 1 teaspoon cinnamon.

Heat oven to 400°. In large pan, combine sugar, cornstarch and salt. Stir in berries, rhubarb, lemon juice and margarine. Cook over medium heat, stirring frequently until hot, bubbly and slightly thickened. Pour into 2-quart casserole. Separate biscuit dough into 10 biscuits; cut each in half with scissors. Arrange cut-side down on hot fruit mixture in circle around edge of casserole. Sprinkle with sugar-cinnamon mixture. Bake 15 to 20 minutes or until biscuits are golden brown. Serve warm with cream or ice cream. Makes 6 to 8 servings.

Spiced Rhubarb Betty

½ cup butter or margarine
3 cups small white-bread cubes, toasted
1¼ pounds rhubarb, trimmed and cut in ½-inch slices (4 cups)
1 cup firmly packed light brown sugar
2 teaspoons grated orange rind
1 teaspoon ground cinnamon
¼ teaspoon ground nutmeg
⅓ cup hot water
Light cream

Melt butter in a small pan, drizzle over toast cubes in a large bowl and toss until butter is absorbed. Measure ½ cupful and set aside. Preheat oven to 350°. Add rhubarb, brown sugar, orange rind, and spices to bread mixture; toss until evenly mixed. Spoon into a 1½ quart baking dish; sprinkle the ½ cup buttered toast cubes on top. Drizzle hot water over mixture; cover. Bake 45 minutes; uncover and bake 15 minutes, or until rhubarb is tender and topping crisp. Cool in dish on a wire rack. Spoon into serving dishes; serve warm with cream. Makes 6 servings.

Lemon Rhubarb Brown Betty

2 cups bread crumbs or graham-cracker crumbs
3 tablespoons melted butter
2½ cups rhubarb
½ cup brown or granulated sugar
1 tablespoon lemon juice
½ teaspoon grated lemon peel
⅓ cup hot water

Combine crumbs and butter; stir over low heat until lightly browned. Place one-third in greased 8"x8"x2" pan. Chop rhubarb; arrange half in layer over crumbs. Sprinkle with half the sugar, lemon juice and peel. Add second layer of crumbs and remaining rhubarb, sugar, lemon juice and peel. Cover with remaining crumbs. Pour water over mixture. Bake in moderate oven (375°) until rhubarb is tender, 30 to 40 minutes. Serve warm. Makes 6 servings.

Rhubarb Ambrosia Betty

5 cups rhubarb (cut in ½-inch pieces)
1¾ cups sugar
1 tablespoon all-purpose flour
¼ teaspoon salt
½ tablespoon freshly grated orange rind
1 orange, peeled, sectioned and diced
4 cups fresh bread cubes
¼ pound butter, melted
½ cup unsweetened dried flaked coconut

Mix together rhubarb, sugar, flour, salt, half the orange rind and the diced orange. Add half the bread cubes and half the butter. Mix well. Transfer to a greased 8-inch square baking pan. Combine remaining orange rind, bread cubes and butter and the coconut. Sprinkle over the top of the rhubarb layer. Place in an oven preheated to 375° and bake for 40 minutes or until browned. Serve warm. Serves 6 to 8. Note: Frozen rhubarb, without syrup, may be substituted for the fresh rhubarb.

Rhubarb Dumplings

> **Dough:**
> 2 cups flour
> 1 teaspoon baking powder
> 2½ tablespoon shortening
> 1 teaspoon salt
> ¾ cup milk
> **Filling:**
> 2 cups rhubarb
> 1 cup sugar
> Nutmeg
> **Topping:**
> 1½ cup water
> 1½ cup sugar
> 1½ tablespoon flour

Mix ingredients and roll out like cinnamon rolls. Place rhubarb on dough and sprinkle sugar and nutmeg over rhubarb. Roll up and cut pieces 1" thick. Place in 9"x13" pan. Boil topping for 3 minutes and pour over dumplings before baking. Bake at 350° for 35 minutes.

Microwave Caramel Rhubarb Dumplings

> ¾ cup packed brown sugar
> ¼ cup granulated sugar
> 3 tablespoons cornstarch
> 3 tablespoons margarine
> 3 cups sliced rhubarb
> 1¼ cups all-purpose flour
> ¼ cup granulated sugar
> 1½ teaspoon baking powder
> ¼ cup margarine
> ⅓ cup milk
> 2 teaspoons granulated sugar
> ¼ teaspoon ground cinnamon

In a two-quart microwave-safe casserole combine 1¼ cups water, the brown sugar, the first ¼ cup granulated sugar, cornstarch, and the 3 tbsp. margarine. Add rhubarb. Cook uncovered on 100% power for 7-9 minutes or till mixture is thickened and bubbly, stirring after every minute. Cover casserole to keep warm. In a bowl combine flour, remaining ¼ cup granulated sugar, baking powder and ¼ teaspoon salt. In a custard cup cook the remaining margarine till melted; add to dry ingredients along with milk. Mix till just blended. Drop batter by rounded tbsp. onto rhubarb mixture. Cover with waxed paper; cook 5-6 minutes or till dumplings are done. Stir together the two tsp. sugar and cinnamon; sprinkle over dumplings. Serve warm.

Strawberry-Rhubarb Grunt

> 1 pint strawberries, hulled and halved
> 1 pound rhubarb, cut into one-inch pieces
> ½ cup sugar
> 3 tablespoons water
> ½ teaspoon ground cinnamon
> ½ teaspoon lemon juice
> 1 cup Bisquick
> ⅓ cup milk
> 1 cup heavy or whipping cream

Combine strawberry halves, rhubarb, sugar, water, cinnamon and lemon juice in a ten-inch skillet. Heat to boiling. Meanwhile, mix Bisquick and milk in a small bowl until just moistened. Drop dough by heaping tablespoons on top of simmering strawberry mixture, forming eight dumplings. Cover and cook the strawberry mixture about 8-10 minutes until dumplings are set. Spoon strawberry mixture and dumplings into dessert bowls. Top with cream.

Cranberry-Rhubarb Swirls

1½ cups sugar
½ cup cranberry juice cocktail
¼ cup water
3½ cup rhubarb, cut in ½-inch pieces
2 cups biscuit mix
½ teaspoon nutmeg
1 tablespoon oil
⅓-½ cup milk
2 tablespoons soft butter or margarine

Combine 1 cup sugar, cranberry juice and water in saucepan. Bring to boil; boil 1 minute. Put half of rhubarb in greased 8"x8"x2" pan. Combine biscuit mix, 2 tablespoons sugar and nutmeg. Add oil and milk to make soft dough; knead lightly on floured surface. Roll into 9" square. Spread with butter; sprinkle with remaining rhubarb and 4 tablespoons sugar. Roll as jelly roll; seal; cut into 9 slices. Place slices, cut side up, over rhubarb in pan. Pour over cranberry syrup; sprinkle with 2 tablespoons sugar. Bake in hot oven (425°) 25-30 minutes. Serve warm. Makes 9 servings.

Bisquick Rhubarb-Strawberry Rolls

2 cups Bisquick
¾ cup milk
2 cups rhubarb
1 cup strawberries
½ cup sugar
1½ cups sugar
2 cups water

Mix the Bisquick and milk. Roll into oblong shape to 1-inch thickness. Mix the rhubarb, strawberries and ½ cup sugar; spread over bisquick mix and roll. Cut onto 1½-inch slices. Mix the 1½ cups sugar and water and boil. Pour into buttered 8"x8" pan. Place the slices in the syrup. Bake at 450° for 25-30 minutes. Serve with whipped cream.

Rhubarb-Strawberry Rolls

Dough
2 cups sifted enriched flour
4 teaspoons baking powder
1 tablespoon sugar
½ teaspoon salt
⅓ cup shortening
1 well-beaten egg
½ cup milk

Sift flour, baking powder, sugar, and salt; cut in shortening until mixture is like coarse crumbs. Combine egg and milk; add to the sifted dry ingredients, stirring just until moistened. Roll the dough out ½-inch thick; brush with melted butter.

Fruit filling
1 cup water
¾ cup sugar
2 cups diced rhubarb
1 cup sliced strawberries

Make syrup of water and sugar; cook 5 minutes. Pour into greased 8"x8"x2" square pan. Sweeten the rhubarb and strawberries to taste and spread onto the shortcake; roll as for jelly roll. Cut ½-inch slices; place, cut side down, in hot syrup in the square pan. Bake in hot oven (425°) for 30 minutes. Serve hot with cream. Makes 8 servings.

Rhubarb Shortcake

½ cup flour
Dash salt
1 tablespoon sugar
1 teaspoon baking powder
1½ teaspoon grated orange peel
2 tablespoons unsalted butter, cut into small pieces
3 tablespoons milk
Rhubarb filling (recipe follows)
Whipped cream or plain yogurt sweetened with a little brown sugar, optional

Stir together flour, salt, sugar, baking powder and orange peel. Cut in butter to make a crumbly mixture. Add milk, a tablespoon at a time, to make a slightly sticky dough. Divide dough in half. Pat into two round ½-inch cakes, 2¾-3 inches in diameter, on lightly floured surface. Place rounds two inches apart on ungreased baking sheet. Bake at 400° until lightly browned, 14-16 minutes. Remove biscuits to wire rack and cool until just warm. Cut each biscuit in half horizontally. Place bottoms of each on two serving plates. Top each with half the rhubarb filling. Top with remaining biscuit halves. Serve plain or with dollops of whipped cream. Makes two servings.

Rhubarb Filling

2 cups rhubarb, sliced ½-inch thick (about 3 long stalks)
½ cup sugar
1 tablespoon flour, optional
1 tablespoon orange liqueur, optional
1 tablespoon butter

Combine rhubarb and sugar in medium, heavy-bottomed pan. Do not add water. Cover and cook over medium heat, stirring occasionally, until rhubarb breaks down and becomes syrupy, about 10 minutes. Stir together flour and liqueur. Stir into rhubarb mixture. Cook over low heat until rhubarb mixture is thick, about one minute. Remove from heat and stir in butter. (Rhubarb will have pleasant consistency even if it isn't thickened with flour. To eliminate flour, just cook rhubarb and sugar, remove from heat and stir in butter.) Makes two servings.

Kuchens & Strudels

Rhubarb Kuchen
Rhubarb Kuchen à la Mode
Frozen Dough Rhubarb Kuchen
Indian Rhubarb Pockets
Rhubarb and Amaretti Strudel

Rhubarb Kuchen

Filling:
3 cups rhubarb, cut up
1½ cups sugar
2 tablespoons cornstarch
¼ cup cold water
1 teaspoon vanilla

Dissolve cornstarch in ¼ cup cold water. Mix with rhubarb, sugar and vanilla. Cook until thick and rhubarb is tender. Set aside and cool.

Crust:
½ cup butter
½ cup flaked coconut
1 box yellow cake mix

Cut butter into dry cake mix until crumbly; stir in coconut. Pat crumb mixture lightly into ungreased 9"x13" pan. Build up edges slightly. Bake in 350° oven for 10 minutes. Remove from oven; arrange cooled rhubarb mixture on warm crust, to within ½ inch of sides of pan.

Topping:
½ cup sugar
1 teaspoon cinnamon
1 cup heavy whipping cream
1 egg

Mix sugar and cinnamon; sprinkle on rhubarb. Blend cream and egg thoroughly; drizzle over rhubarb. Bake in 350° oven for 35 minutes or until edges are light brown and cream mixture is lightly set. Serve warm or cold. Makes 12 three-inch squares.

Rhubarb Kuchen à la Mode

1 cup flour
1 tablespoon sugar
1½ teaspoon baking powder
½ teaspoon salt
2 tablespoons butter
1 egg
2 tablespoons milk

1 3-oz. package strawberry gelatin
⅓ cup sugar
3 tablespoon flour
5 cups rhubarb
⅔ cup sugar
⅓ cup flour
3 tablespoons butter
Ice cream

Combine the 1 cup flour, 1 tablespoon sugar, baking soda and salt. Cut in 2 tablespoons butter until mixture is coarse crumbs. Beat egg with milk; add to flour mixture. Stir until dry ingredients are moistened. Pat dough evenly on bottom of 9"x9"x2" baking pan. Combine the gelatin, the ⅓ cup of sugar and 3 tablespoons flour. Add to rhubarb and mix well, then turn into crust-lined pan. Combine remaining ⅔ cup sugar, and ⅓ cup flour, cut in remaining butter until crumbly. Sprinkle evenly over rhubarb filling. Bake in 350° oven for 45 minutes or till tender and topping is lightly brown. Cool. Cut in 8" squares and top with ice cream. Serves eight.

Frozen Dough Rhubarb Kuchen

1 loaf frozen white or sweet bread, thawed
2 cups rhubarb, thinly sliced
2 tablespoons margarine
½ cup sugar
1 teaspoon cinnamon
1 egg
¼ cup cream

Let dough rise slightly. Press into well-greased 9"x13" pan. Arrange rhubarb on top. Cut butter into sugar and cinnamon. Sprinkle over rhubarb. Cover; let rise until double in size (1-1½ hour). Bake in 350° oven for 20-25 minutes until light golden brown. Beat egg and cream together until well blended. Spoon over rhubarb. Bake 10-15 minutes more until custard is set and rhubarb is tender.

Indian Rhubarb Pockets

1½ cups flour
½ teaspoon salt
¼ cup water
1 teaspoon baking soda
6 tablespoons shortening
   Rhubarb, diced and sweetened with honey

Combine flour, baking soda and salt. Add the shortening, cutting it in with a knife until the consistency is that of fine meal. Add the water slowly until a stiff dough forms. Roll out dough mixture on floured board to ½" thickness. Cut dough into 4" rounds. Place a large tablespoon of rhubarb in center. Top with second flour round and seal by fluting edges. Bake in 400° oven for 10 minutes or until golden brown. Drizzle with vanilla glaze.

Rhubarb and Amaretti Strudel

1 pound fresh rhubarb stalks
9 sheets phyllo dough (about half a package), thawed if frozen
1 cup unsalted butter, melted
1 cup crushed amaretti cookies
½ cup packed brown sugar
2 teaspoons ground cinnamon
   Grated zest of 1 orange
½ cup chopped walnuts

Cut the rhubarb into ½" pieces. Blanch the pieces in boiling water for 1½ minutes. Drain and pat dry. Place 1 sheet of the phyllo on a baking sheet the same size as the phyllo dough. Brush with butter and sprinkle with a few tbsp. of the amaretti crumbs. Keep the remaining dough covered with a damp towel to prevent it from drying out. Layer 4 more sheets of the phyllo on the top of the first in the same manner, brushing each with butter and sprinkling with crumbs. Then layer 2 more sheets, brushing with butter but omitting the amaretti. Preheat oven to 375°. Make a compact row of the rhubarb pieces 2" from one long edge of the dough. Sprinkle the rhubarb with the brown sugar, cinnamon, orange zest, walnuts, and any remaining amaretti crumbs. Layer 2 more sheets of the phyllo dough, brushing with butter, over the rhubarb and last sheet of phyllo. Starting at the edge with the rhubarb, roll up the dough like a jelly roll. Turn seam side up and brush the top and sides of the dough with butter. Turn seam side down, fold both ends under to seal, and brush the top with butter. Bake 40-50 minutes. Let stand several minutes and then slice. Serve with whipped cream. Makes 8 portions.

Crisps & Crunches

Rhubarb Crisp
Almond-Rhubarb Crisp
Microwave Rhubarb Custard Crisp
Granola Rhubarb Crisp
Blueberry-Rhubarb Crisp
Fresh Strawberry-Rhubarb Crisp
Fresh Strawberry-Rhubarb Crisp (Microwave)
Rhubarb Crunch
Oatmeal Rhubarb Crunch
Rhubarb-Strawberry Crunch

Rhubarb Crisp

    3 cups cut up rhubarb
    ¾ cup sugar
    1 egg
    3 tablespoons flour
    ⅓ cup brown sugar
    ¾ cup flour
    4 tablespoons butter

Scald rhubarb and drain. Mix with sugar, flour, and egg. Place in 8"x12" baking dish. Combine topping, spread on top and bake 45 minutes at 350°.

Almond-Rhubarb Crisp

    1½ pounds fresh rhubarb
    ¾ cup granulated sugar
    2 tablespoons flour
    ¼ teaspoon mace
    ⅛ teaspoon salt
      Few drops red food coloring
      (optional)
    ¼ cup water
    1 egg, beaten
    ⅓ cup brown sugar
    ½ cup flour
    ¼ cup butter
    ½ cup almonds, roasted, diced

Cut rhubarb into ¾-inch pieces to make 5 cups. Mix sugar, 2 tablespoons flour, mace and salt. Add food coloring to water. Add to rhubarb with sugar mixture and egg, toss gently. Turn into 8- or 9" square pan. Mix brown sugar, flour and butter to crumbly state, stir in almonds. Sprinkle over rhubarb. Bake at 375° for 30 minutes. Serve warm or cold and if desired, with cream. Serves 8.

Microwave Rhubarb Custard Crisp

    3 cups sliced rhubarb, cut ½" thick
    ½-⅔ cup granulated sugar
    3 tablespoons flour
    ¼ teaspoon nutmeg
    ¼ cup half-and-half milk
    2 eggs
    ⅓ cup margarine, softened
    ½ cup flour
    ½ cup oatmeal (not instant)
    ¼ cup brown sugar
    ¼ cup chopped nuts
    ½ teaspoon cinnamon

Combine rhubarb, sugar, 3 tbsp. flour and nutmeg in a microwavable 9" or 8"x8" dish. Add the half-and-half and eggs; mix well and set aside. Combine margarine, flour, oatmeal, brown sugar, nuts and cinnamon; mix with fork or pastry blender until crumbly. Sprinkle evenly over rhubarb mixture. Microwave on high 12-14 minutes, rotating dish as necessary for even cooking, until rhubarb in center is tender. Serve warm or cold.

Granola Rhubarb Crisp

    4 cups rhubarb, cut ¼-inch thick
    ¾ cup granulated sugar
    2 tablespoons all-purpose flour
    ⅓ cup softened margarine or butter
    ⅓ cup firmly packed brown sugar
    1 teaspoon vanilla extract
    ¾ cup oatmeal (not instant)
    ½ cup flour
    ¼ cup coconut
    ¼ cup sunflower nuts or chopped walnuts
    ½ teaspoon ground cinnamon
      Vanilla ice milk or whipped cream

Combine rhubarb, granulated sugar and 2 tablespoons flour in microwave-safe 8"x8" dish; stir well and set aside. In separate bowl, stir together margarine, brown sugar and vanilla until smooth. Pour remaining ingredients atop margarine mixture. Using a fork, mix all ingredients until crumbly. (Mixture should not form a dough.) Stir rhubarb again to make sure sugar is well mixed. Top with crumbly mixture. Microwave on high for 10-12 minutes or until rhubarb is tender when tested with fork. If necessary for even cooking, rotate dish once or twice. Serve warm or cold, with or without ice milk or whipped cream. Makes 9 servings.

Blueberry-Rhubarb Crisp

2½ cups rhubarb, cut ½-inch thick
⅓ cup granulated sugar
2 tablespoons all-purpose flour
1 can (20oz.) blueberry pie filling
½ cup margarine
¾ cup wheat or all-purpose flour
¾ cup oatmeal (not instant)
⅓ cup packed brown sugar
¾ teaspoon cinnamon

Combine rhubarb, sugar and 2 tablespoons all-purpose flour in a deep microwavable dish (8"x8"); stir in blueberry pie filling and set aside. Combine margarine, whole wheat flour, oatmeal, brown sugar and cinnamon; mix with fork or pastry blender until crumbly. Sprinkle evenly over rhubarb-blueberry mixture. Microwave on high 14-17 minutes, rotating dish as necessary for even cooking, until center starts to bubble and rhubarb is tender. Serve warm or cold. Makes 8-12 servings.

Fresh Strawberry-Rhubarb Crisp

6 stalks rhubarb
1½ pints strawberries
6 tablespoons sugar
2 tablespoons all-purpose flour
Almond Topping
1 cup slivered almonds
1 cup all-purpose flour
1 cup packed brown sugar
⅛ teaspoon salt
8 tablespoons unsalted butter, cold
For Serving
Vanilla ice cream, or sweetened whipped cream

Preheat oven to 350°. Toast almonds in the preheated oven until lightly browned (6-8) minutes. Cool and chop coarsely. Combine one cup flour, brown sugar, salt and almonds in a mixing bowl. Cut butter into ¼-inch pieces and add to the flour mixture. Rub the mixture in the palm of your hands, incorporating the butter while keeping the topping crumbly. Set aside in a cool place. Peel away the thin red outer skin of the rhubarb and trim off the ends. Cut the stalks into ¾ to 1-inch pieces; you should have about three cups. Hull the strawberries and cut them in half. Toss the fruit together with the granulated sugar and two tablespoons of flour until evenly coated. Place the fruit in a 6"x10" baking dish and cover evenly with topping. Bake for one hour or until brown, bubbly and crisp. Serve warm. Dish into glass bowls and serve with ice cream or whipped cream.

Fresh Strawberry-Rhubarb Crisp (Microwave)

4 cups rhubarb, cut up
1½ cups strawberries, sliced
1 cup sugar
1 tablespoon lemon juice
1 tablespoon quick-cooking tapioca

Topping:
⅔ cup flour
½ cup quick-cooking rolled oats
¼ cup chopped nuts
3 tablespoons brown sugar
½ teaspoon cinnamon
¼ teaspoon nutmeg
¼ cup margarine

Combine rhubarb, strawberries, sugar, lemon juice and tapioca in 8"x8"x2" dish. Combine flour, oats, nuts, brown sugar and spices in mixing bowl. Cut in margarine with pastry blender. Sprinkle on top of fruit mixture. Microwave on full power for 13-15 minutes or until rhubarb is tender. Turn dish once or twice during cooking time. May be served warm or cold with whipping cream or vanilla ice cream.

Rhubarb Crunch

1 cup sifted flour
½ cup butter
5 tablespoons powdered sugar
2 eggs
¼ cup flour
   Pinch of salt
1½ cups sugar
¼ teaspoon baking powder
2 cups unsweetened, cut rhubarb

Mix together first three ingredients and pat in the bottom of 8" or 9" square pan. Bake in moderate oven 350° for 15 minutes or until golden brown. Beat eggs until fluffy and gradually add sugar, flour, baking powder and salt which have been sifted together. Add rhubarb and pour over crust. Bake again in moderate oven for 35 minutes longer. Variation: One can of drained pie cherries and ½ cup of chopped nuts may be substituted for the rhubarb. Excellent when served with whipped cream.

Oatmeal Rhubarb Crunch

1 cup sifted flour
¾ cup uncooked oatmeal
⅓ cup brown sugar, firmly packed
½ cup melted butter
1 teaspoon cinnamon
4 cups diced rhubarb
1 cup sugar
2 tablespoons cornstarch
1 cup water
1 teaspoon vanilla

Mix flour, oatmeal, brown sugar, melted butter and cinnamon until crumbly. Press half of mixture into greased 9" baking pan. Cover crumb mixture with diced rhubarb. Combine sugar, cornstarch, water and vanilla. Cook until thick. Pour this sauce over rhubarb. Top with remaining crumbs and bake at 350° for one hour. Serve warm, plain or topped with whipped cream.

Rhubarb-Strawberry Crunch

Crust and topping:
2 cups flour
2 cups packed brown sugar
1 cup butter or margarine
2 cups rolled oats, uncooked

Filling:
6 cups rhubarb in ½-inch slices
1 cup granulated sugar
2 tablespoons cornstarch
1 cup cold water
1 teaspoon vanilla
1 (3 ounce) package strawberry-flavored gelatin

Mix flour and brown sugar in a large bowl. Cut in butter as for pie crust. Mix in rolled oats. Gently pat half the mixture into bottom of a 9"x13" pan. Save remaining half for topping. Spoon cut-up rhubarb evenly over the crust. Mix granulated sugar and cornstarch in a saucepan. Stir in water and cook over medium heat until mixture comes to a boil and is clear and thickened. Add vanilla. Spoon over rhubarb. Sprinkle dry gelatin over the top. Top with remaining crumbs. Bake at 350° for 40-45 minutes. Serve either warm or cold and topped with whipped cream or ice cream, if desired. Makes 12 to 15 servings.

Bars, Cookies & Candy

Rhubarb Bars
Rhubarb Oatmeal Squares
Rhubarb Dream Bars
Rhubarb-Filled Bars
Rhubarb Fruit Squares
Rhubarb Cream Cheese Bars
Rhubarb Candy
Orange Candy Slice Rhubarb Bars
Rhubarb-Squash-Sour Cream Bars
Rhubarb Cookies
Spiced Rhubarb Cookies
Rhubarb Filling
Rhubarb-Filled Cookies
Rhubarb-Filled Chocolate Cookies
Crunchy Drops
Thumbprint Cookies
Rhubarb Pizza

Rhubarb Bars

1 cup flour
1 teaspoon baking powder
¼ teaspoon salt
¼ cup butter
1 egg
1 tablespoon milk
2 cups cut-up rhubarb
1 3 oz. package strawberry gelatin
¼ cup butter
1 cup sugar
½ cup flour

Combine 1 cup flour, baking powder and salt. Cut in ¼ cup butter. Add egg and milk, mix and pat into a greased, 9" pan. Cover with rhubarb and sprinkle with dry gelatin. Mix remaining butter, sugar and flour and put on rhubarb. Bake 50 minutes at 350°. Cut into squares.

Rhubarb Oatmeal Squares

3 cups rhubarb (cut up)
1½ cups sugar
2 tablespoons cornstarch
1 teaspoon vanilla
¼ cup water

Dissolve cornstarch in water, add rhubarb and sugar. Cook slowly until thick. Add vanilla and cool. Mix:

1½ cups quick-cooking oatmeal
1½ cups flour
1 cup brown sugar
½ teaspoon soda
1 cup shortening

Add ½ cup chopped nuts. Put ¾ of this mixture into pan. Pour on cooled rhubarb mixture and cover with remaining crumbs. Bake at 375° for 25-30 minutes.

Rhubarb Dream Bars

Crust:
2 cups flour
½ cup powdered sugar
1 cup butter or margarine

Topping:
4 eggs
2 cups sugar
½ cup flour
¾ teaspoon salt
4 cups diced rhubarb

For crust mix together the flour, powdered sugar and butter. Press into a ungreased 9"x13" pan. Bake 15 minutes in moderately slow oven (325°). For topping, beat eggs with mixer, add sugar and beat well. Add the flour and salt. Mix well. Fold in diced rhubarb. Spread over baked crust. Sprinkle with cinnamon or nutmeg, if desired. Return to oven for 45 minutes at 325°. Cut into squares.

Rhubarb-Filled Bars

3 cups rhubarb, cut in ½" slices
1 cup sugar
1 teaspoon orange flavoring
½ teaspoon cinnamon

Mix the above ingredients in bowl and let stand. Mix the following:

½ teaspoon salt
2½ cups flour
1½ teaspoon soda
1 cup brown sugar
⅔ cup butter

Mix until butter is evenly distributed. Add 1½ cup dry oatmeal. Pat ⅔ into a 9"x13" buttered pan. Spread with rhu-

Bars, Cookies & Candy

73

barb mixture. Sprinkle remaining oatmeal mixture over rhubarb. Bake at 400° for 25 minutes or until browned. Remove from oven and when cool, cut into bars.

Rhubarb Fruit Squares

- ⅔ cup molasses
- 2 cups each diced rhubarb and granola
- ⅛ teaspoon salt
- ½ cup cooking oil
- 1 cup whole wheat flour

Blend molasses and oil, then add granola, flour and salt. Press one-half the mixture in 7"x11" baking pan. Pour rhubarb over mixture, layer with remaining mixture. Bake at 400° for 25 minutes. Cool and cut. (Hint: cook rhubarb till hardly tender before adding to cut tartness.)

Rhubarb Cream Cheese Bars

- ½ cup margarine
- 1½ cup flour
- ½ cup brown sugar
- ½ cup oatmeal
- ½ teaspoon salt
- 1½ cup rhubarb, cut up
- 8 ounces cream cheese, softened
- 1 egg
- ½ teaspoon nutmeg
- ¼ teaspoon cinnamon
- ¾ cup sugar

Mix margarine, flour, oatmeal, brown sugar and salt until crumbly. Press half of this mixture into an 8"x8" pan. Save half for top. Beat cream cheese, sugar, egg, nutmeg and cinnamon until smooth. Stir in rhubarb. Pour over crust. Top with remaining crust mixture. Pat gently. Bake in 350° oven for 40-50 minutes. Store in refrigerator.

Rhubarb Candy

- 2 cups rhubarb sauce, sweetened
- 2 tablespoons unflavored gelatin
- ½ cup cold water
- 1⅓ cup sugar
- 1 tablespoon cornstarch
  Dash of salt
- ⅔ cup chopped walnuts
- 1 teaspoon grated lemon peel
- 1 tablespoon lemon juice
- ⅓ cup confectioner's sugar (for rolling)

Soften gelatin in cold water. Mix sugar, cornstarch and salt. Add rhubarb pulp. Cook again over low heat, stirring constantly, until thick. Add gelatin; stir until gelatin until dissolved and mixture is again thick. Remove from heat. Stir in walnuts, lemon peel and juice. Turn into a shallow 8" glass dish that has been rinsed in cold water. Let stand 24 hours. Cut in rectangles, about 50. Roll in confectioner's sugar and place on rack until dry. Store in covered container. Makes one pound.

Orange Candy Slice Rhubarb Bars

    1 eight-ounce package refrigerated
       crescent rolls
    ⅔ cup brown sugar
    ⅔ cup butter
    2 cups rhubarb in ½" pieces
    ⅔ cup orange candy slices, cut up
    1 cup nuts, chopped

Unroll the rolls, but do not separate them. Pat rolls flat in 9"x13" pan. Set aside. Melt brown sugar and butter in a pan. Then add rhubarb and candy slices; bring to a boil and boil one minute. Pour over roll dough in pan. Sprinkle with chopped nuts. Bake in 375° oven for 15-20 minutes. Makes 12 three-inch bars.

Rhubarb-Squash-Sour Cream Bars

    ¾ cup brown sugar
    ¼ cup margarine
    1 egg
    1 teaspoon vanilla
    1 cup whole wheat flour
    ½ teaspoon baking soda
    ¼ teaspoon salt
    ½ cup sour cream
    ½ teaspoon lemon juice
    ½ cup rhubarb, diced
    ½ cup summer squash, diced
    ¼ cup sugar
    ½ cup walnuts, chopped
    1 tablespoon margarine
    ½ teaspoon cinnamon

Cream brown sugar and margarine. Add egg and vanilla. Stir well. Mix flour, soda and salt; then stir into sugar mixture, alternating with sour cream. Add lemon juice; stir well. Add rhubarb and squash; stir well. Pour into greased and floured

8"x8" pan. Mix topping—sugar, margarine and cinnamon. Stir in nuts. Sprinkle over batter. Bake in 350° oven for 45 minutes. Cool in pan. Makes 9 servings.

Rhubarb Cookies

    1½ cup brown sugar
     2 eggs
     1 cup shortening
     ½ teaspoon salt
     3 cups flour
     1 teaspoon soda, mixed with 3
        tablespoons boiling water
    1½ cups rhubarb, finely cut
        You may use ¾ cup chocolate
        chips, coconut or raisins

Mix sugar, shortening, eggs and soda. Add flour, salt and rhubarb. Mix well with spoon. Bake at 350° for about 10 minutes. May frost with powdered sugar, a little butter, cream and vanilla.

Spiced Rhubarb Cookies

    1½ cup rhubarb, diced
     1 cup raisins or dates
     ¼ cup water
     ½ cup shortening
    1½ cups sugar
     2 eggs
     ½ teaspoon salt
    2½ cups flour
     1 teaspoon cinnamon
     1 teaspoon baking soda
     1 cup nuts

Cook first three ingredients until soft. Add remaining ingredients in same pan. Drop by big teaspoon on greased cookie sheet. Bake at 350° for 15 minutes.

Rhubarb Filling (for 2 following recipes)

- 2 cups powdered sugar
- 3 tablespoons light cream
- ⅛ teaspoon salt
  Sweetened rhubarb sauce
  A few drops red food coloring

Mix sugar, salt and cream together. Add rhubarb sauce (see *Sauces & Soups* section) to make a soft filling. Stir in a few drops of food coloring. Put cookies together in pairs with the filling.

Rhubarb-Filled Cookies

- 1½ cup flour
- ½ cup butter
- 1 cup sugar
- 1 tablespoon milk
- 2 eggs
- 1 teaspoon baking powder
- 1 teaspoon vanilla
- ½ teaspoon salt

Cream butter; add sugar and cream well. Add milk alternately with flour, baking powder and salt. Chill dough. Roll ½-inch thickness on floured board. Cut; add rhubarb filling and cover with another cookie. Bake in 400° oven 10 minutes. Makes 2 dozen.

Rhubarb-Filled Chocolate Cookies

- ⅔ cup margarine
- 1 cup sugar
- 1 egg, beaten
- 2 cups flour
- 1 teaspoon baking powder
- ½ teaspoon soda
- ½ teaspoon salt
- ¾ cup cocoa, unsweetened
- ¼ cup milk

Cream margarine and sugar until fluffy. Beat in egg. Sift flour, baking powder, soda, salt and cocoa into bowl. Add milk and mix to a soft dough. Form the dough into a two-inch roll and refrigerate overnight. Cut ⅛-inch slices and place on ungreased cookie sheet. Bake in 325° oven for 10-12 minutes or until firm to touch. Cool on wire rack.

Crunchy Drops

- 1 cup margarine
- 1¼ cup brown sugar
- 2 eggs
- 1 teaspoon vanilla
- 1 cup cooked rhubarb sauce, sweetened and drained
- 2 cups flour
- ½ teaspoon soda
- ½ teaspoon salt
- ⅛ teaspoon mace or nutmeg
- 2 cups quick cooking oatmeal
- 1 cup walnuts, chopped

Cream margarine and brown sugar. Beat in eggs and vanilla. Stir in rhubarb. Mix flour, soda, salt and mace. Add to sugar mixture and stir well. Stir in oatmeal and nuts. Drop by teaspoonful on greased cookie sheet. Bake in 350° oven for 12-13 minutes or until lightly browned. Makes four dozen.

Thumbprint Cookies

⅔ cup sugar
2 egg yolks
1 cup butter
2 teaspoons vanilla
2 cups flour
½ teaspoon salt
   Rhubarb jam (see *Jams & Jellies* section)

Cream together sugar and butter. Add one egg yolk, cream and add other yolk and cream. Add vanilla, flour and salt. Mix and roll into balls the size of a walnut. Make a dent in center with thumb and put jam in dent. Bake at 375° for 15 minutes.

Rhubarb Pizza

¼ cup sugar
1 cup flour
1 teaspoon baking powder
¼ teaspoon salt
2 tablespoons shortening
1 egg, beaten with 2 tbsp. milk
3 cups rhubarb
1 box strawberry gelatin
½ cup powdered sugar
½ cup sugar
½ cup flour
½ cup margarine

Mix ¼ cup sugar, 1 cup flour, baking powder, salt, shortening and egg. Line bottom and sides of round pizza pan with mixture. Cover with rhubarb and sprinkle gelatin over top. Mix powdered sugar, ½ cup sugar, remaining flour and melted margarine; sprinkle over rhubarb and gelatin. Bake at 350° for 45 minutes.

Desserts

Betty's Rhubarb Dessert
Quick Rhubarb Dessert
Rhubarb Peekaboo
Royal Rhubarb Dessert
Rhubarb Cherry Dessert
Rhubarb Cream Dessert
Marshmallow Rhubarb Dessert
Rhubarb Crunch Dessert
Rhubarb Delight Dessert
Rhubarb Dream Dessert
Golden Rhubarb Dessert
Orange Rhubarb Dessert
Refrigerator Rhubarb Dessert
Rhubarb Meringue
Rhubarb Meringue Dessert
Cranberry-Rhubarb Dessert
Rhubarb-Pretzel Dessert
Rhubarb Cream Cheese Dessert
Baked Rhubarb Sponge
Raspberry-Rhubarb Shortcake
Frozen Rhubarb Cream

Betty's Rhubarb Dessert

⅓ cup sugar
5½ cups rhubarb
1 package strawberry gelatin
1 small package white cake mix
⅔ cup water

Mix the first three ingredients together. Combine the cake mix and water. Spread over the rhubarb mixture. Dot butter over the top. Bake at 350°. Cool after baking.

Quick Rhubarb Dessert

4 cups rhubarb
1 cup sugar (can be omitted)
2 small boxes of strawberry gelatin
1 package yellow or white cake mix
1 cup cold water
⅓ cup melted butter or margarine
1 greased 9"x13" glass baking dish

Add first four ingredients in above order. Pour water and butter over this. Bake 1 hour at 350°. Easy.

Rhubarb Peekaboo

1 package two-layer-size white or yellow cake mix
Egg and oil for preparing cake mix
4 cups diced rhubarb
1 teaspoon grated lemon rind
1 cup granulated sugar

Grease a 9"x13" baking dish. Heat oven to 350°. Prepare cake batter as directed on package. Spread cake batter in prepared pan. Top with rhubarb and grated rind; sprinkle sugar over rhubarb. Bake 40 to 45 minutes. As the dessert bakes, the rhubarb goes to the bottom, forming a sauce, cake rises to the top. If desired, sprinkle top of cake with powdered sugar right after baking. Serve warm with cream. Makes 12 servings.

Royal Rhubarb Dessert

Crust:
1 cup sifted flour
1 teaspoon baking powder
⅓ teaspoon salt
2 tablespoons butter
1 beaten egg
2 tablespoons milk

Filling:
4 cups fresh rhubarb, cut up
1 package strawberry gelatin

Topping:
⅔ cup sugar
½ cup flour
⅓ cup butter

Combine flour, baking powder, salt and butter. Mix as for pie crust, add beaten egg and milk. Pat in bottom and sides of a 9"x9" baking pan. Arrange cut up rhubarb in crust. Sprinkle the dry gelatin over evenly. Mix the sugar, flour and butter together and sprinkle over dry gelatin. Bake in oven 350°, 30 to 40 minutes. Cool, cut into squares to serve. Top with whipped cream or ice cream.

Rhubarb Cherry Dessert

Crust:
½ pound margarine
2 cups flour
2 tablespoons sugar

Crumb together with pastry cutter and spread in 9"x13" pan. Bake in 350° oven for 10 minutes.

Filling:
5 cups rhubarb, cut up
1 cup sour cherries
5 egg yolks
2 cups sugar
4 tablespoons flour
¼ teaspoons salt
1 cup cream

Mix ingredients together and pour on crust. Bake in 350° oven for 40-45 minutes.

Topping:
5 egg whites
8 tablespoons sugar
2 teaspoons vanilla
Dash salt
Chopped nuts

Beat egg whites until stiff. Add sugar, one tablespoon at a time, beating well after each addition. Mix in vanilla and salt. Spread on custard filling and sprinkle with chopped nuts or coconut. Bake in 350° oven for 10-15 minutes or until golden brown.

Rhubarb Cream Dessert

2 eggs
1 cup sugar
1 cup sweet cream
1 teaspoon vanilla
1 teaspoon salt
2 cups flour
2 teaspoon baking powder
5 cups rhubarb (cut up)
1½ cups sugar

Beat the eggs and add sugar gradually, alternately with sweet cream. Add vanilla, salt, flour, and baking powder. Beat well. Pour half of batter into ungreased 9"x13" pan. Mix rhubarb and sugar. Pour this over batter, then pour remaining batter on top. Bake for 45 minutes at 350°.

Marshmallow Rhubarb Dessert

2 cups crushed graham crackers
¼ cup sugar
½ cup butter
4 cups rhubarb
2 cups sugar
1 package red gelatin
2 cups miniature marshmallows
1 cup whipped cream

Melt butter in pan. Add graham crackers and sugar. Press into bottom of 8"x12" pan. Cook rhubarb slightly. Add sugar and gelatin. Cool to lukewarm. Add marshmallows. Let stand. Add whipped cream and pour into pan. Chill until set.

Rhubarb Crunch Dessert

Crust:
1 cup cake flour
½ cup butter
5 tablespoons powdered sugar

Filling:
1½ cups sugar
2 eggs, well-beaten
½ teaspoon salt
⅓ cup flour
¾ cup nuts
¾ teaspoon baking powder
2 cups diced rhubarb

Mix as pie crust: cake flour, butter and powdered sugar and pat into an 8" pie plate. Bake at 350° for 15 minutes. Mix remaining ingredients. Blend and pour on top of crust and return to oven to bake 30 minutes more. Serve with ice cream or whipped cream.

Rhubarb Delight Dessert

1 small package orange gelatin
1 cup hot water
1 cup rhubarb (cooked, sweetened, canned or freshly cooked)
⅓ cup mayonnaise
4 ounces cream cheese
¼ teaspoon lemon juice
½ pint whipping cream

Mix and leave thicken: gelatin, water and rhubarb. Combine and beat until smooth: mayonnaise, cream cheese and lemon juice. Whip whipping cream. To whipped cream add thickened gelatin and mayonnaise mixture. Beat until thoroughly mixed. Chill for 3 hours before serving. Serve in sherbet glasses with a half cherry on top. Recipe makes 4 cups or eight ½-cup servings. This dessert freezes very well.

Rhubarb Dream Dessert

Crust:
1 cup sifted flour
5 tablespoons powdered sugar
½ cup butter or margarine

Topping:
2 beaten eggs
1½ cups sugar
¼ cup flour
¾ teaspoon salt
2 cups finely chopped rhubarb

Blend crust ingredients together. Press into an ungreased 7½"x11½" pan and bake in a 350° oven 15 minutes. Mix topping ingredients together. Spoon over the crust and bake at 350° for 35 minutes. Serve warm with whipped or plain cream. Serves 6.

Golden Rhubarb Dessert

Crust:
1 cup flour
2 tablespoons sugar
½ cup butter
Pinch of salt

Filling:
4 cups diced rhubarb
3 beaten egg yolks
1½ cups sugar
¾ cup cream or rich milk
4 tablespoons flour

Mix crust: flour, sugar, butter and salt and pat into 8x11" pan. Bake at 325° for 10 minutes. Mix filling ingredients and pour over baked crust. Bake for 45 minutes. Make a meringue of 3 egg whites, adding 5 tablespoons sugar. Spread on top of baked dessert and return to oven to brown.

Orange Rhubarb Dessert

Crust:
2 cups all-purpose flour
2 teaspoons sugar
1 cup butter or margarine

Filling:
1½ cups sugar
2 tablespoons all-purpose flour
5 cups sliced rhubarb
½ cup milk
¼ cup fresh orange juice

Meringue:
3 eggs, separated
¼ cup plus 2 tablespoons sugar

Combine flour and sugar; cut in butter with a pastry blender until mixture resembles coarse meal. Press crumb mixture evenly in bottom of a 13"x9"x2" baking pan. Bake at 375° for 15 to 20 minutes or until crust is browned. Set aside. For filling, combine sugar and flour in a heavy 5-quart saucepan. Add rhubarb, milk, and orange juice. Cook over medium heat, stirring constantly, until thickened. Beat egg yolks at high speed of an electric mixer until thick and lemon colored. Gradually stir about ¼ of hot mixture into yolks; add to remaining hot mixture, stirring constantly. Cook over medium heat 2 to 3 minutes, stirring constantly. Pour over crust. Cool. Beat egg whites (at room temperature) until

foamy; gradually add sugar, 1 tablespoon at a time, beating 2 to 4 minutes or until stiff peaks form and sugar dissolves. Spread meringue over filling. Bake at 350° for 12 to 15 minutes or until browned. Yield: 12 servings.

Refrigerator Rhubarb Dessert

20 (1 package) crushed graham crackers
¼ cup butter
4 cups chopped rhubarb
1 cup sugar
½ cup water (if using frozen rhubarb, decrease water to 2 tablespoons)
3 tablespoons cornstarch
1 cup whipping cream
1 tablespoon sugar
1½ cups miniature marshmallows.
1 small package instant vanilla pudding

Mix graham crackers and butter and pat into a pan saving a little to sprinkle over top of dessert. Combine rhubarb, sugar, water and cornstarch; bring to a boil and simmer 2 to 3 minutes until thickened. Spoon thickened sauce over crust — cool. Whip whipping cream and sugar and fold in miniature marshmallows. Pour over sauce. Prepare instant vanilla pudding as directed. Spoon over cream mixture. Sprinkle remaining crumbs, refrigerate overnight.

Rhubarb Meringue

1 cup cake flour
5 tablespoons powdered sugar
½ cup butter
3 eggs
2 cups sugar
½ cup flour
1 teaspoon baking powder
3 cups rhubarb, cut up

Beat together 1 cup cake flour, powdered sugar, and butter. Put in 9"x13" pan and bake at 350° for 15 minutes. Beat the eggs, sugar, flour, baking powder, and rhubarb. Spread on top of first mixture and bake at 350° for 45 minutes. Top with whipped cream.

Rhubarb Meringue Dessert

Crust:
1 cup butter
2 tablespoons powdered sugar
2 cups flour

Custard:
5 cups rhubarb
6 egg yolks, beaten
2 cups sugar
4 tablespoons flour
½ teaspoon salt

Meringue:
6 egg whites
Pinch of cream of tartar
1 teaspoon vanilla
12 tablespoons sugar
½ teaspoon salt

Mix and pat 9"x13" pan with butter, powdered sugar and flour. Bake at 350° for 15 minutes. Mix and stir for the custard rhubarb, egg yolks, sugar, flour, and salt. Pour into crust and bake 45 minutes or until done. Whip egg whites to peaks, add cream of tartar, vanilla, sugar and salt. Spread and bake until light and brown.

Cranberry-Rhubarb Dessert

2 cups flour, sifted
½ cup brown sugar, firmly packed
2 teaspoons baking powder
¼ teaspoon salt
½ cup shortening
2 eggs
¾ cup milk
6 cups rhubarb, cut up
1 one-pound can whole cranberry sauce
2 tablespoons quick-cooking tapioca
1 3-oz. package raspberry gelatin
1 cup sugar
½ cup flour
5 tablespoons butter
½ cup nuts, chopped (optional)

Combine 2 cups flour, brown sugar, baking powder and salt in a bowl. Cut in shortening until crumbly. In a separate bowl beat eggs with milk. Add to flour mixture; blend well. Spread in 9"x13" greased pan. In a separate bowl, combine rhubarb, cranberry sauce and tapioca. Spoon over dough in pan, leaving about ½-inch dough around edge. In a separate bowl combine gelatin, sugar and ½ cup flour. Cut in butter until mixture is crumbly. Sprinkle over rhubarb; top with nuts. Bake in 350° oven 50-60 minutes, serve warm or cold.

Rhubarb-Pretzel Dessert

Crust:
2½ cups crushed pretzels
¾ cup melted butter
2 tablespoons sugar

Mix together and reserve ½ cup for topping. Press remainder in 9"x9" pan; bake in 350° oven for 10 minutes. Cool.

Filling:
1 cup sugar
½ cup water
3 tablespoons cornstarch
4 cups rhubarb, cut up

Combine sugar, cornstarch and water; stir in rhubarb. Cook over low heat until thickened. Continue to cook about five minutes more until rhubarb is tender. If desired, add a couple drops of red food coloring. Cool. Spread into pretzel crust.

Topping:
½ cup whipping cream, whipped *or*
1 cup whipped topping
1½ cup miniature marshmallows

Mix and spread over rhubarb mixture.

Rhubarb Cream Cheese Dessert

Crust:
1 cup flour
¼ cup granulated sugar
½ cup butter or margarine

Rhubarb layer:
3 cups fresh rhubarb, cut in ½-inch pieces
½ cup sugar
1 tablespoon flour

Cream layer:
12 ounces cream cheese, softened
½ cup sugar
2 eggs

Topping:
8 ounces sour cream
2 tablespoons sugar
1 teaspoon vanilla

For crust, mix flour, sugar and butter; pat into 10" pie plate. Set aside. For rhubarb layer, combine rhubarb, sugar and flour; toss lightly and pour into crust. Bake at 375° for 15 minutes. Meanwhile, prepare cream layer by beating together cream cheese and sugar until fluffy. Beat in eggs one at a time, then pour over hot rhubarb layer. Bake at 350° for about 30 minutes or until almost set. Combine topping ingredients: spread over hot layers. Chill. Yield: 12-16 servings.

Baked Rhubarb Sponge

3 cups rhubarb cut into ½" pieces
1½ teaspoons quick-cooking tapioca
1 cup sugar
1 egg
½ cup sifted flour
1 teaspoon baking powder
⅛ teaspoon salt
2 tablespoons orange juice

Combine rhubarb, ⅔ cup sugar and tapioca. Put in six individual custard cups or one-quart casserole. Beat egg until thick; gradually beat in remaining sugar. Sift dry ingredients together and add to egg mixture alternately with orange juice. Pour over rhubarb. Bake in a moderate oven at 350° for 30 minutes if in custard cups, or 45 minutes if in casserole.

Frozen Rhubarb Cream

4 cups finely chopped rhubarb
¾-1 cup sugar
1 teaspoon grated orange peel
½ cup orange juice
½ cup light corn syrup
2 egg whites
2 tablespoons sugar
1 cup whipping cream, whipped
   Toasted coconut or nuts, if
   desired
   Orange liqueur, if desired

In large saucepan, combine rhubarb, ¾ to 1 cup sugar, orange peel, juice and corn syrup. Heat to boiling, stirring constantly. Reduce heat and cook until rhubarb is tender, about 10 minutes; cool. Pour mixture into loaf pan or freezer trays; freeze to slush consistency, about 1 to 2 hours. Beat egg whites until frothy; gradually add 2 tablespoons sugar, beating constantly until stiff peaks form. Spoon partially frozen rhubarb mixture into large bowl; beat until light and frothy. Fold beaten egg whites and whipped cream into rhubarb mixture. Return to freezer container; cover with plastic wrap. Freeze until firm, about 3 to 4 hours, stirring occasionally. Allow to soften slightly, spoon into sherbet glasses. Garnish with toasted coconut or nuts; drizzle with orange liqueur. Makes 12 (½-cup) servings. Tip: To toast coconut or nuts, bake at 350° for 5 to 7 minutes or until light brown, stirring once.

Raspberry-Rhubarb Shortcake

1 12-oz. package frozen raspberries
2 cups rhubarb, cut
¾ cup sugar
   Bisquick box mix

Cook raspberries and rhubarb with sugar over medium heat (about 10 minutes). Stir occasionally. Cool to room temperature. Make Bisquick shortcake by recipe on box. Roll out dough ¾-inch thick; cut into rounds. Bake 10-12 minutes at 425°. Split biscuits in half; add raspberry-rhubarb mixture by tablespoons. Top with vanilla ice cream or yogurt.

Cakes & Tortes

Rhubarb Coffee Cake

Rhubarb Spice Cake

Rhubarb Bisquick Coffee Cake

Jane's Recipe for Coconut Rhubarb Cake

Rhubarb Bran Cake

Rhubarb Carrot Cake

Rhubarb Pudding Cake

Rhubarb-Chocolate Cake

Rhubarb Cocoa Cake

Rhubarb-Molasses Shortcake

Vanilla Rhubarb Cake

Sour Cream Rhubarb Cake

Rhubarb Upside Down Cake

Rhubarb Cream Cake

Rhubarb Marshmallow Upside-Down Cake

Red-Hot Candy Upside Down Cake

Rhubarb Cheesecake

Rosy Rhubarb Torte

Gelatin Rhubarb Upside-Down Cake

Bisquick Rhubarb Torte

Rhubarb Custard Torte

Harriet's Rhubarb Torte

Rhubarb Coffee Cake

- ½ cup shortening
- 1½ cups brown sugar
- 1 egg
- 1 teaspoon vanilla
- 1 cup sour milk
- 2 cups flour
- 1 teaspoon soda
- ¼ teaspoon salt
- 2 cups diced uncooked rhubarb
  Topping:
- ½ cup nut meats
- ½ cup brown sugar
- 1 teaspoon flour

Cream the shortening and the 1½ cups brown sugar, add the egg, well beaten, and the vanilla, stir in the sour milk and blend well. Sift the flour, soda and salt together, add to the creamed mixture. Stir in rhubarb. Combine topping ingredients; sprinkle on top of cake. Bake for 35 to 40 minutes in a 350° oven.

Rhubarb Spice Cake

- 1½ cups flour
- ¼ teaspoon salt
- 1 cup sugar (can reduce sugar)
- 1 teaspoon soda
- 1 egg, beaten
- 1 cup sour cream
- 1 teaspoon vanilla
- 2 cups rhubarb, cut fine
- ½ cup brown sugar
- ½ cup nuts
- ¼ teaspoon cinnamon

Mix first 8 ingredients. Place in 9"x13" pan and sprinkle with a topping made from remaining ingredients. Bake at 350° for about 45 minutes.

Rhubarb Bisquick Coffee Cake

- 2¼ cups Bisquick
- 1 cup brown sugar
- 1 egg
- 1 cup milk
- 1 teaspoon vanilla or almond flavoring
- 1¼ cup rhubarb
- ½ cup sliced almonds
- ½ cup sugar
- 1 tablespoon margarine
- ¼ cup sliced almonds

Heat oven to 350°. Grease and flour jellyroll pan, 15"x10"x1". Mix Bisquick, brown sugar, egg, milk and flavoring; beat. Stir in rhubarb and ½ cup almonds; pour into pan. Mix sugar, margarine and ¼ cup almonds. Sprinkle evenly over batter in pan. Bake 25-30 minutes.

Jane's Recipe for Coconut Rhubarb Cake

- 1 cup butter
- 1½ cup brown sugar
- 1 egg
- 2 cups flour
- ½ teaspoon salt
- 1 teaspoon soda
- ½ cup milk
- 2 teaspoon vanilla
- 2 cups cut fresh rhubarb

  Topping:
- 1 (7 ounce) package coconut
- 1 teaspoon cinnamon
- ½ cup sugar

Cream butter and sugar in large mixing bowl. Add egg, beat well. Sift flour with salt and soda. Blend in dry ingredients alternately with milk and vanilla. Stir in rhubarb. Turn into greased 9"x13" pan. Sprinkle coconut, sugar and cinnamon over batter. Bake at 350° for 40 minutes.

Rhubarb Bran Cake

- 2 cups sliced rhubarb, 1-inch lengths
- 3 tablespoons butter or margarine
- 1½ cups sugar
- ½ cup bran flakes
- ¾ cup flour
- 1 teaspoon baking powder
- ½ teaspoon salt
- ½ cup milk
- 2½ teaspoons cornstarch
- ½ cup boiling water

Put rhubarb into a greased, 8" square baking dish. Cream together the butter and ¾ cup of the sugar; stir in the bran until well blended. Sift the flour; measure; sift again with the baking powder and ¼ teaspoon of the salt. Stir into the creamed mixture alternately with the milk until well blended. Spread the batter evenly over the rhubarb. Stir the remaining ¾ cup sugar, ¼ teaspoon salt and the cornstarch until blended. Sprinkle over batter. Pour boiling water over cake and bake in a moderately hot oven (375°) for about 50 minutes or until browned. Makes eight servings.

Rhubarb Carrot Cake

- 2 cups flour
- 2 teaspoons baking powder
- 1½ teaspoons salt
- 2 teaspoons cinnamon
- 2 cups sugar
- 1½ teaspoons baking soda
- 1½ cups oil
- 4 eggs
- 2 cups grated carrots
- 1 cup unsalted walnuts
- 2 cups sweetened rhubarb*

Sift together the flour, baking powder, salt, soda and cinnamon. Add sugar, oil and eggs. Stir in the carrots, walnuts and drained rhubarb. Bake in greased and floured 13"x9"x2" pan 45-50 minutes at 350°. * To sweeten rhubarb, place cut rhubarb in baking dish, sprinkle with 1 cup sugar and cover with foil. Bake in 350° oven for ½ hour or until soft.

Rhubarb Pudding Cake

- 4 cups diced fresh rhubarb
- 1 cup sugar
- ¾ cup water
- ¼ cup shortening
- ½ cup sugar
- 1 egg
- ½ teaspoon vanilla
- 1 cup sifted flour
- 2 teaspoon baking powder
- ¼ teaspoon salt
- ½ cup milk

Cook rhubarb, 1 cup sugar and water until rhubarb is tender; keep hot. Cream shortening and ½ cup sugar, beat in egg and vanilla. Sift flour, baking powder and salt together and add alternately with milk to creamed mixture. Pour batter into a greased 9"x9" baking pan. Spoon hot rhubarb sauce over batter. Bake in a 350° oven for 40 minutes. The batter rises to the top and the sauce sinks to the bottom. Ladle into dishes and serve with heavy cream, or slightly melted ice cream. Cut into 9 squares, or servings.

Rhubarb-Chocolate Cake

Beat well:
1½ cup sugar
2 eggs
½ cup shortening
⅓ cup cocoa
Add:
½ cup cold milk
2 cups + 2 tablespoons flour
1½ teaspoons soda
¼ teaspoon salt
Beat in:
1 cup cold water
¾ cup cooked, sweetened rhubarb
½ teaspoon vanilla
1 teaspoon almond extract

Bake in greased 9"x13" pan at 350° for 35 minutes.

Rhubarb Cocoa Cake

⅔ cup butter
1½ cups sugar
3 eggs
1 teaspoon vanilla
½ cup unsweetened cocoa
2¼ cups sifted flour
1 teaspoon baking powder
¾ teaspoon baking soda
¼ teaspoon salt
1 cup rhubarb juice (see *Beverages* section)
1 cup minced rhubarb, drained

Cream butter and sugar. Beat in eggs and vanilla. Sift dry ingredients and add alternately with juice to creamed mixture. Stir in rhubarb. Pour batter into greased and floured bundt pan. Bake at 325° for 55-60 minutes. Allow to cool 15 minutes before removing from pan.

Rhubarb-Molasses Shortcake

2 cups flour
2 teaspoons baking soda
1 teaspoon ginger
1 teaspoon cinnamon
1 teaspoon nutmeg
½ teaspoon salt
¾ cup brown sugar
1 cup sour milk or buttermilk
¾ cup melted shortening
¾ cup light molasses
2 beaten eggs

Sift dry ingredients, less spices if you prefer a milder shortcake. In large bowl, stir in brown sugar. Combine milk, eggs, shortening and molasses; add to flour mix. Beat smooth. Pour into greased 9"x9" pan. Bake at 350° for 40-45 minutes. Serve warm. Cut cake in two layers. Spoon rhubarb sauce between layers and on top. Top with whipped cream.

Rhubarb Sauce:
4 cups fresh rhubarb
1 cup sugar
Dash of salt
¾ cup water
2 tablespoons cornstarch
¼ cup water

Cook rhubarb, sugar, salt and ¾ cup water. Cook 10 minutes. Blend cornstarch and ¼ cup water. Stir into rhubarb. Cook, stirring, until mixture boils. Cook two minutes.

Vanilla Rhubarb Cake

Cake:
1 cup granulated sugar
1 egg
2 tablespoons melted butter or margarine
1 cup buttermilk or sour milk
½ teaspoon salt
½ teaspoon baking soda
1 teaspoon baking powder
2 cups all-purpose flour
1 cup diced fresh rhubarb

Topping:
2 tablespoons melted margarine
½ cup granulated sugar

Vanilla sauce:
½-1 cup granulated sugar
½ cup margarine
½ cup evaporated milk
1 teaspoon vanilla

Blend together sugar, egg and butter. Beat in buttermilk until smooth. Stir together salt, baking soda, baking powder and flour. Stir dry ingredients into buttermilk mixture; mix well. Stir in rhubarb. Pour into a greased 9" square baking pan. Combine topping ingredients; sprinkle on top of batter. Bake at 350° for 45 minutes or until cake tests done. For sauce, mix sugar, margarine and milk; bring to boil and cook 1 minute, stirring constantly. Remove from heat; stir in vanilla. Serve sauce over cake. Yield: 12 servings.

Sour Cream Rhubarb Cake

1½ cup brown sugar
½ cup melted butter
1 egg
2 cups flour
1 teaspoon baking soda
¼ teaspoon salt
1 cup sour cream
1½ cup rhubarb, cut into small pieces

Cream together the brown sugar and butter. Add the egg. Sift together the flour, baking soda and salt and mix with the above. Add the sour cream and rhubarb pieces. Bake in a greased and floured 13"x9" pan at 350° for 45-50 minutes.

Rhubarb Upside Down Cake

3 cups rhubarb
2 cups flour
½ teaspoon salt
1 teaspoon soda
½ cup shortening
¾ cup sugar
2 eggs
1 teaspoon vanilla
½ cup sour milk
Sugar, to sprinkle on top

Put rhubarb (cut in pieces) into a bowl and put sugar with it. Set aside. Sift dry ingredients together in bowl. Make a well in the flour mixture and add shortening and eggs, vanilla and sour milk. Beat until smooth. Put rhubarb in a 9"x9" square pan and pour batter over the rhubarb. Sprinkle with sugar and bake 45-50 minutes at 350° Serve warm.

Rhubarb Cream Cake

1 package yellow cake mix (no
   pudding in the mix)
4 cups sliced rhubarb
1 cup sugar
½ pint whipping cream

Mix box cake according to directions and spread in greased 9"x13" cake pan. Sprinkle with the rhubarb, then the sugar and finally, slowly pour the whipping cream over all. Last three ingredients will form a custard and sink through the cake. Bake at 350° for 40-45 minutes. After baking, turn upside down onto a cookie sheet or metal tray. Optional: Serve with ice cream or whipped cream. Easy.

Rhubarb Marshmallow Upside-Down Cake

4 cups rhubarb, cut fine
1 bag miniature marshmallows
   Yellow or white cake mix
   Cool Whip

In bottom of 9"x13" greased pan, put rhubarb, cut fine, and marshmallows. Pour the cake mix, prepared according to directions on box, over the rhubarb. Bake at 350° for 30 to 40 minutes. Serve warm, upside down, with Cool Whip.

Red-Hot Candy Upside Down Cake

3 cups rhubarb, chopped
10 large marshmallows, quartered
¾ cup sugar
12 red hots
¼ cup butter
1 cup sugar

2 eggs, beaten
1¾ cup flour
3 teaspoons baking powder
¼ teaspoon salt
¼ cup milk

Arrange rhubarb in greased 9"x13" pan. Add quartered marshmallows and ¾ cup sugar. Sprinkle with red hots. Thoroughly cream butter with 1 cup sugar. Add eggs and beat well. Add sifted dry ingredients alternately with milk. Pour batter over rhubarb. Bake in 350° oven one hour. Cool and turn onto plate. Serve with whipped cream or ice cream. Makes 12-15 servings.

Rhubarb Cheesecake

Crust:
1 cup oatmeal
½ teaspoon cinnamon
¼ cup melted butter
¾ cup finely chopped walnuts
2 tablespoons brown sugar

Filling:
3 cups cooked rhubarb
2 8-oz. packages cream cheese
1 cup sugar
1 teaspoon vanilla
1 tablespoon water
4 eggs
1½ tablespoons cornstarch
½ cup sour cream

Combine crust ingredients until well blended. Press into bottom and sides of 10" springform pan. In mixing bowl or food processor blend until smooth the cream cheese, sugar, sour cream, vanilla, eggs and two cups of the rhubarb. Pour over crust. Bake at 325° for one hour.

Turn off oven. Allow cake to cook in oven with door ajar for an additional hour. Meanwhile, mix cornstarch and water with remaining cup of rhubarb. Cook until it thickens and turns clear. Spread mixture over cake and refrigerate.

Rosy Rhubarb Torte

Crust:
1 cup flour
2 tablespoons sugar
1 cup graham cracker crumbs
½ cup melted butter
½ cup chopped nuts

Filling:
3 cups diced rhubarb
1 3-oz. pkg. strawberry gelatin
½ cup sugar
1 8-oz. package cream cheese
1¾ cup powdered sugar
¼ package Dream Whip

Mix first 5 ingredients together and pat in 9"x13" ungreased pan. Bake at 350° for 10 minutes. Cook rhubarb with sugar; remove from heat and add gelatin. Cool in refrigerator until it thickens. Cream the cream cheese and powdered sugar. Whip Dream Whip as on directions; fold in cream cheese mix. Then spread ½ of this on crust, then the rhubarb mixture, ending with another layer of the cream cheese mixture. Makes 18 servings.

Gelatin Rhubarb Upside-Down Cake

5 cups cut-up rhubarb
1 3-oz. pkg strawberry gelatin
1 cup sugar
3 cups miniature marshmallows
1 package white cake mix

Put rhubarb in bottom of 9"x13" pan. Sprinkle with gelatin. Then sprinkle with sugar and top with marshmallows. Prepare cake mix according to package directions. Put on top of marshmallows. Bake at 350° for 1 hour. Cool and remove onto foil tray or leave in pan and put in refrigerator.

Bisquick Rhubarb Torte

Crust:
1½ cups Bisquick
3 tablespoons melted butter
¼ cup milk
Filling:
4½ cups cut-up rhubarb
1 3-oz. pkg. strawberry or raspberry gelatin
Streusel:
1½ cups sugar
½ cup margarine
¾ cup flour

Mix crust ingredients well and pat in greased 7"x11" pan. Cover with rhubarb. Sprinkle with strawberry or raspberry gelatin. Mix streusel ingredients well and sprinkle over filling. Bake for 45 minutes at 375°.

Rhubarb Custard Torte

Crust:
1 cup shortening
2 cups flour
3 tablespoons sugar
2 egg yolks
1 teaspoon vanilla
Dash of salt

Custard:
2 tablespoons melted butter
1 large can evaporated milk
4 cups rhubarb
2 cups sugar
4 tablespoons flour
4 egg yolks

Meringue:
¼ teaspoon cream of tartar
6 egg whites
⅓ cup sugar

Mix the crust ingredients and pat into a 9"x13" pan. Mix the custard ingredients well and pour into the crust. If desired, a small amount of tapioca may be put into the crust before adding custard. Bake at 375° for 45 minutes. Add cream of tartar to egg whites and beat until very stiff. Add a little vanilla and the sugar a small amount at a time. Remove torte from oven and cover with meringue. Bake until meringue browns slightly. Makes 20 servings.

Harriet's Rhubarb Torte

Crust:
1 cup flour
2 tablespoons sugar
1 teaspoon salt
½ cup butter or margarine

Filling:
2 cups cut up rhubarb
2 tablespoons flour
1 grated orange rind
3 egg yolks
2 cups sugar

Meringue:
3 egg whites
6 tablespoons sugar
⅓ cup cream or half and half

Mix crust like pie crust and pat in 9"x13" pan and bake 15 minutes at 350°. Mix rhubarb, flour and rind of orange and put on top of crust. Separate 3 eggs, beat yolks, add sugar, put on top of rhubarb. Bake for approximately 50 minutes. Beat egg whites, add sugar 2 tablespoons at a time, add cream or half and half. Put on baked rhubarb and brown in 350° oven.

Pies

No-egg Rhubarb Pie

Jeanne's No-egg Rhubarb Pie

Cooked Rhubarb No-egg Pie

Spiced Rhubarb No-egg Pie

Grandma Hildie's No-egg
Rhubarb Pie

Tart Cherry Rhubarb Pie

Rhubarb Cherry Pie

Cranberry Rhubarb Pie

Juneberry-Rhubarb Pie

Rhubarb-Mulberry Pie

Peach-Rhubarb Pie

Raspberry-Rhubarb Pie

Rhubarb Refrigerator Pie

Sandcherry-Rhubarb Pie

Deep Dish Rhubarb Pie with
Cream Cheese Pastry

Rhubarb and Strawberry
Bavarian Pie

Rhubarb-Strawberry-Yogurt Pie

Rhubarb Chiffon Pie

Eleanor's Rhubarb Pie

Top Crust Rhubarb Custard Pie

Easy-Crust Rhubarb Custard Pie

Rhubarb No-Crust Custard Pie

Sweet Cream Rhubarb Custard Pie

Rhubarb Sour Cream Custard Pie

Rhubarb Meringue Pie

Rhubarb-Orange Custard Pie

Rhubarb and Raisin Pie

Rhubarb-Pineapple Pie

Honey Strawberry-Rhubarb Pie

Rhubarb-Pineapple Custard Pie

Honey Lemon Rhubarb Pie

Rhubarb-Strawberry-Orange Pie

Rhubarb Strawberry Streusel Pie

Glazed Strawberry-Rhubarb Pie

Irresistible Strawberry-Rhubarb Pie

Cooked Strawberry-Rhubarb Pie

Strawberry and Rhubarb Pie

No-egg Rhubarb Pie

1½ cups sugar
¼ teaspoon salt
5 tbsp. flour or 3 tbsp. tapioca
2 tablespoons grated orange rind
4 cups chopped rhubarb

Combine first four ingredients. Place half of rhubarb in pastry lined pan. Sprinkle with half of sugar mixture. Repeat. Dot with butter and cover with crust. May substitute 1 cup crushed pineapple for 1 cup rhubarb.

Jeanne's No-egg Rhubarb Pie

Double pie crust
4-5 cups rhubarb, cut-up
1 3-oz. box of strawberry gelatin
½ tablespoon Minute tapioca
1 cup of sugar

Pour boiling water over rhubarb. Let set 5 minutes. Drain. Pour in bottom pie shell. Sprinkle ½ of the box of strawberry gelatin over rhubarb. Sprinkle ½ tablespoon tapioca over pie. Add 1 cup of sugar. Dot with margarine. Add upper crust. Bake at 425° about 35-40 minutes.

Cooked Rhubarb No-egg Pie

Baked 9" pie shell
2½ cups sugar, approximately
1⅓ cups water
1¾ pounds rhubarb (5 cups cut)
6 tablespoons cornstarch
1 cup heavy cream, whipped

Cool the pie shell while preparing the filling. Mix the sugar and one cup water and bring to a boil, stirring. Add the rhubarb and cook, stirring gently once or twice before the rhubarb starts to soften. Continue cooking without stirring until just tender. Remove the rhubarb with a slotted spoon and reserve. Reheat the syrup to boiling. Blend the cornstarch with the remaining water and add to the boiling syrup. Cook, stirring, until thickened and clear. If desired, add more sugar. Add the reserved rhubarb, cool five minutes and turn into the pie shell. Chill. Before serving, garnish with whipped cream, sweetened if desired.

Spiced Rhubarb No-egg Pie

3 cups chopped rhubarb
1½ cups sugar
1 tablespoons lemon juice
2 tablespoons flour
½ teaspoon cinnamon
¼ teaspoon ground cloves
Pastry for 2-crust 9" pie
2 tablespoons butter or margarine

Combine rhubarb, sugar, lemon juice, flour, cinnamon and cloves and blend well. Spoon into pie shell and dot with butter. Top with remaining pastry, sealing and fluting edges. Slash top in 5 or 6 places. Bake at 400° 20 minutes. Reduce heat and bake at 350° 20 to 30 minutes longer, until pastry is golden brown. Makes 6 to 8 servings.

Grandma Hildie's No-egg Rhubarb Pie

1 cup half and half or cream
1 generous cup cut-up rhubarb
2 tablespoons flour
1 cup sugar

Mix sugar and flour together; add cream

and rhubarb. Put in unbaked 8-inch pie shell. Do not cover with a top crust. Bake at 450° for 15 minutes, then turn down to 350° and bake 30-45 more minutes. Makes 6 servings. Easy.

Tart Cherry Rhubarb Pie

    2 cups sliced rhubarb
    1 21-oz. cherry pie filling
    ½ cup sugar
    2½ tablespoons quick-cooking tapioca
       Pastry for two-crust lattice top pie

Hint: Use fluted pastry wheel to cut lattice crust.

Combine sliced rhubarb, cherry pie filling, sugar and tapioca. Let mixture stand 15 minutes. Line nine-inch pie plate with pastry. Fill with rhubarb mixture. Carefully adjust top lattice crust; seal and flute edges. Bake in 400° oven for 40-45 minutes.

Rhubarb Cherry Pie

    3 cups rhubarb
    1 can tart red cherries (2 cups), drained
    1¼ cup sugar
    ¼ cup quick-cooking tapioca
    5 drops red food coloring
    2 pastry shells

Line 9" pie pan with pastry. Combine sugar, cherries, rhubarb and tapioca. Mix well. Let stand 15 minutes. Add coloring and put in pie shell. Add top crust; flute edge. Bake at 400° for 40-50 minutes.

Cranberry Rhubarb Pie

    Combine and let stand:
    2½ cups cranberries (cooked)
    3 tablespoons water
    1½ cups finely cut rhubarb
    3 tablespoons quick-cooking tapioca

Prepare two-crusted 9" pie pan. Pour in ingredients and adjust top crust. Bake at 400° for 45-60 minutes or until rhubarb is tender.

Juneberry-Rhubarb Pie

       Pastry for two-crust, 9" pie
    1½ cup sugar
    3 tablespoons flour
    1 egg, beaten
    1 tablespoon lemon juice
    2 tablespoons melted butter
    3 cups cubed rhubarb
    1½ cup juneberries

Mix sugar, flour, beaten egg, lemon juice and melted butter. Add rhubarb and juneberries; mix well and let stand while making crusts. Pour rhubarb filling into unbaked pie shell. Add top crust and seal edges. Bake in 350° oven for approximately one hour, or until filling is bubbling in the middle.

Rhubarb-Mulberry Pie

  2 cups sliced rhubarb
  1 cup mulberries
  1 cup sugar
     Dough for a two-crust pie
  ¼ teaspoon nutmeg
  4 tablespoons flour
  2 tablespoons butter

Combine rhubarb and mulberries in bowl. Combine sugar, nutmeg and flour. Sprinkle about ½ of the mixture in the bottom of a 9" pie pan lined with pastry. Top with rhubarb-mulberry mixture and add remaining sugar-flour mixture. Dot with butter. Add top crust, cut vents and sprinkle with cinnamon and sugar (about 2 tsp. sugar and ¼ tsp. cinnamon). Bake at 425° for 45-50 minutes until crust is browned and juices are bubbling.

Peach-Rhubarb Pie

  1 15-oz. package refrigerated pie crust
  1 teaspoon flour
  ¾-1 cup sugar
  2 12-oz. packages frozen peaches, thawed and drained, reserve juice
  3 cups fresh diced rhubarb
  ⅓ cup flour

Prepare pie crust as directed for two-crust pie. Preheat oven to 350°. In saucepan, combine sugar and one teaspoon flour. Gradually add 1 cup reserved fruit liquid. Cook until mixture boils and thickens, stirring constantly. Shake rhubarb and peaches in the ⅓ cup flour. Place flour-coated fruits in pie-crust-lined pan. Pour thickened juice over fruit. Top with second crust and flute edges. Make slits in top crust for vents. Bake 50-60 minutes or until golden brown.

Raspberry-Rhubarb Pie

     Combine:
  1 cup sugar
  ¼ teaspoon nutmeg
  ¼ cup orange juice
  1 cup raspberries
  ¼ teaspoon salt
  2 tbsp. quick-cooking tapioca
  3 cups rhubarb
  1 double-crust pastry

Place ingredients in 9" pastry-lined pie tin. Top crust with crimped edges. Bake at 450° for 10 minutes, then 30 minutes at 350°.

Rhubarb Refrigerator Pie

  1 package orange gelatin
  ½ cup sugar
  1 cup heavy cream, whipped
     red food coloring (optional)
  1 9" baked pastry shell
  1 cup hot water
  ¼ cup cold water
  2½ cup rhubarb
  ⅓ cup water

Dissolve gelatin in 1 cup hot water and add ¼ cup water; cool. Cook rhubarb with ⅓ cup water until tender; measure 1½ cups rhubarb and sugar, cooking until sugar dissolves. Cool. When gelatin is partially thickened, beat until light. Fold in rhubarb and then whipped cream. If needed, add food coloring. Chill until mixture begins to mound. Pile into pastry shell and chill until firm.

Sandcherry-Rhubarb Pie

Pastry for two-crust, 9" pie
1½ cup pitted sandcherries (fresh or frozen)
1½ cups diced rhubarb
1½ cup sugar
1 tablespoon butter
1 tablespoon flour

Wash and pit fresh sandcherries. Dice fresh rhubarb. Thaw and drain frozen fruit. Mix fruit, sugar and flour. Place in pastry-lined pie plate. Dot filling with butter; cover with pastry. Bake in 425° oven for 40 minutes or until filling bubbles and pastry is nicely browned.

Deep Dish Rhubarb Pie with Cream Cheese Pastry

6 cups rhubarb, cut into ½" pieces
½ cup flour
1 cup sugar
1 cup white corn syrup
1 tablespoon butter
3 ounces softened cream cheese
6 tablespoons softened butter
¾ cup flour
½ teaspoon salt

Toss rhubarb with ½ cup flour and turn into 8" square baking pan. Mix sugar and corn syrup. Bring to a boil and pour over rhubarb; dot with butter. Mix cream cheese with butter until fluffy. Add flour and salt and mix with fork. Chill, if necessary, and roll between waxed paper so it is 1 inch larger than baking pan. Place pastry over rhubarb, turning under one inch and crimping against edge. Bake at 425° until pastry is nicely browned. Serve warm with whipped cream.

Rhubarb and Strawberry Bavarian Pie

1 cup sugar
1 pound rhubarb, one-inch pieces
1 9" pastry shell, baked
2 tablespoons unflavored gelatin
1 cup mashed strawberries
1 cup heavy cream, whipped
2 egg whites, stiffly beaten
Whole strawberries

Sprinkle ¾ cup sugar over rhubarb and let stand to extract juice while making pie crust. Soften gelatin in ¼ cup cold water. Heat rhubarb to boiling; lower heat, and simmer until just tender, five minutes or longer. Stir gently or shake pan to prevent sticking. Add softened gelatin and stir gently until dissolved. If desired, remove the most perfect pieces of rhubarb for garnishing and put a spoonful of syrup over them. Set them aside but do not chill. Add strawberries to rhubarb mixture and more sugar, if desired. Cool and then chill until beginning to set. Fold in half the whipped cream and the meringue made by beating remaining sugar into beaten egg whites. Turn into pie shell and chill until set. At serving time, garnish with reserved whipped cream, reserved pieces of rhubarb, and additional whole strawberries. Serves 6-8.

Rhubarb-Strawberry-Yogurt Pie

3 ounce package strawberry gelatin
1¼ cup boiling water
1 eight-ounce size berry yogurt
¼ cup honey
½ pint strawberries
½ pint rhubarb
Baked crust for 9" pie

Combine gelatin and water till dissolved. Beat yogurt and honey into mixture. Chill till partially set, then whip 1-2 minutes till fluffy. Add sliced berries and diced rhubarb folding in easily. Save a few berries for garnish. Pour into cooked crust.

Rhubarb Chiffon Pie

1 tablespoon unflavored gelatin
¼ cup water
3 cups fresh rhubarb
⅓ cup sugar
½ teaspoon salt
½ cup water
2 eggs, separated
A few drops red food coloring (if you are using green rhubarb)
¼ cup sugar
½ cup whipping cream
1 9" oatmeal coconut pie crust (below)

Soften gelatin in ¼ cup water. Combine rhubarb, salt, ⅓ cup sugar and ½ cup water. Cook until rhubarb is tender (about 10 minutes). Add gelatin; stir to dissolve. Beat egg yolks. Stir a little hot rhubarb into egg. Add to rhubarb mixture. Cook 1 minute. Add coloring. Chill slightly. Beat egg whites stiff; fold egg whites and whipped cream into rhubarb mixture. Put in crust; chill until firm.

Oatmeal-Coconut Crust:
1 cup rolled oats
½ cup flaked coconut
⅓ cup soft butter
¼ cup brown sugar

Place oats in shallow pan. Toast in oven at 375° for 7 minutes. Mix well with coconut, butter and brown sugar. Press into 9" pan; chill. Add filling and chill .

Eleanor's Rhubarb Pie

7-9 stalks rhubarb
Boiling water
2 crusts
1 egg
1 cup sugar
2 tablespoons flour
2 tablespoons butter

Cut 7, 8, or 9 stalks up small. Pour boiling water over rhubarb and let stand 2 minutes; drain. Make 2 crusts. Beat 1 egg and pour on bottom crust. In another bowl put 1 cup sugar and 2 tablespoons flour; mix. Pour under and over rhubarb in crust. Dot butter to amount 2 tablespoons. Bake at 425° for 35 minutes.

Top Crust Rhubarb Custard Pie

3 cups cut-up rhubarb
2 egg yolks
1 cup sugar
1 heaping tablespoon flour
1 tablespoon butter
2 egg whites
¼ cup sugar

Mix together rhubarb, egg yolks, sugar, flour and butter. Beat egg whites, adding ¼ cup sugar. Fold egg whites into rhubarb mixture. Put into unbaked pastry crust and bake 10 minutes at 450°, then 45 minutes at 375°. The egg whites will form a top crust on this pie.

Easy-Crust Rhubarb Custard Pie

1 cup flour
5 tablespoons confectioner's sugar
½ cup butter or margarine
2 eggs
1½ cups sugar
¼ teaspoon salt
¼ cup flour
2 cups 2" sliced) rhubarb

Mix flour and confectioner's sugar. Cut in butter with a pastry blender. Press mixture into a 9" pie pan. Bake 15 minutes in a moderate oven (350°). Beat the eggs and stir in sugar, salt and flour. Mix in rhubarb and spoon into hot crust. Bake an additional 35 to 40 minutes. Serve slightly warm or cold. Makes 6 to 8 servings.

Rhubarb No-Crust Custard Pie

4 cups cut-up rhubarb
¾ cup granulated sugar
2 tablespoons butter or margarine
1 cup flour
1 cup granulated sugar
2 beaten eggs
1 teaspoon vanilla
1 tablespoon lemon juice

Grease a 9- or 10" pie tin. Sprinkle with a little flour. Spoon rhubarb into pie tin and sprinkle the ¾ cup of sugar over all. Cut butter into the cup of flour as for pie crust. Add the 1 cup sugar and mix well. Beat eggs and add vanilla. Stir eggs into

the butter-flour mixture. Batter is very stiff. Spoon on top of rhubarb, distributing mixture as evenly as possible. Sprinkle lemon juice on top. Bake at 325° for 1 hour. Serve warm or cold. Makes 6 to 8 servings.

Sweet Cream Rhubarb Custard Pie

Prepare a baked pie shell.

Filling:
2 cups rhubarb, finely sliced
1 cup sugar (may be increased to 1¼ cups if desired)
4 egg yolks, slightly beaten
¼ cup flour (early spring rhubarb may need more)
¼ cup sweet cream (may use 1 tablespoon butter + 3 tablespoons milk)
¼ teaspoon salt

Combine all ingredients and cook over medium/low heat in heavy pan, stirring often to prevent burning. Cook until filling starts to thicken. Set aside to cool, then pour into baked pie shell. Top with meringue.

Meringue:
4 egg whites
8 tablespoons sugar

Beat egg whites until foamy. Gradually add sugar until meringue forms stiff peaks. Pile on cooled pie filling, sealing the meringue to the edges of the pie shell. Brown at 350° for 15-20 minutes until light brown.

Rhubarb Sour Cream Custard Pie

1¼ cups finely chopped rhubarb
1½ cups sugar
2 egg yolks and 1 whole egg
1 cup sour cream
2 egg whites
2 tablespoons sugar

Put rhubarb in pie crust. Mix 1½ cups sugar, eggs except whites of 2, and sour cream. Mix and pour over rhubarb. Bake 1 hour in a 350° oven. Cool slightly. Make a meringue from the 2 egg whites and 2 tablespoons sugar. Return to oven and bake to a golden brown.

Rhubarb Meringue Pie

2 tablespoons butter or margarine
2½ cups cut rhubarb
1 cup sugar
¼ cup sugar
2 tablespoons cornstarch
¼ cup milk
Dash salt
2 beaten egg yolks
2 egg whites
1 8-inch baked pie shell
1 tablespoon sugar

Melt butter; add rhubarb. Add 1 cup sugar and cook until tender. Add ¼ cup sugar, cornstarch, milk and salt. Add beaten egg yolks. Stir rapidly. Cook until thick. Pour in pie shell. Beat egg whites until peaks form. Add 1 tablespoon sugar. Spread on pie and brown in oven.

Rhubarb-Orange Custard Pie

Pastry for 9" pie shell
3 eggs, separated
1¼ cup sugar
¼ cup soft butter or margarine
3 tablespoons frozen orange juice concentrate
¼ cup flour
¼ teaspoon salt
2½ cups rhubarb, cut in ½" pieces
⅓ cup chopped pecans

Line 9" pie pan with pastry; make high fluted rim. Beat egg whites until stiff; add ¼ cup sugar gradually, beating well after each addition. Add butter and juice concentrate to egg yolks; beat thoroughly. Add remaining 1 cup sugar, flour and salt; beat well. Add rhubarb to yolk mixture; stir well. Gently fold in meringue. Pour into pastry-lined pan; sprinkle with nuts. Bake on bottom rack in moderate oven (375°) 15 minutes. Reduce heat to slow (325°); bake 45 to 50 minutes more.

Rhubarb and Raisin Pie

1½ cups rhubarb
1 cup sugar
1 egg
2 tablespoons flour
½ cup seedless raisins

Cut stalks of rhubarb in half-inch pieces before measuring. Mix sugar, flour, and egg; add to rhubarb. Sprinkle the raisins, cut in halves, over the rhubarb and bake between crusts, at 450° for 10 minutes, then 350° for 40 to 45 minutes. Many prefer to scald rhubarb before using; if so prepared, it loses some of its acidity and less sugar is required.

Rhubarb-Pineapple Pie

    4 cups rhubarb sliced
1-1¼ cups sugar
    1 tablespoon quick-cooking
       tapioca
    1 cup crushed pineapple

Combine sugar, rhubarb and tapioca. Fill pie shell with rhubarb, sprinkle with pineapple, adjust top crust and seal the edge of the pie. Bake in a hot oven 30 to 40 minutes.

Honey Strawberry-Rhubarb Pie

3½ cups frozen strawberries, partially
       thawed
    3 cups frozen rhubarb pieces,
       partially thawed
    ¾ cup honey
1½ teaspoons lemon juice
    ¼ cup arrowroot
       Whole wheat pastry for two-
       crust 9" pie

In a medium-size bowl, gently stir together the strawberries, rhubarb, honey, lemon juice and arrowroot. Cook over medium heat, stirring until heated through and thickened. Set aside. Line a 9" pie pan with half of the pastry. Spoon filling into pie shell. Carefully top with remaining pastry and crimp or flute edge. Prick top crust with fork. Bake at 375° for 60 to 65 minutes until filling bubbles. If top crust browns too quickly, cover with foil. Yield: one 9" pie.

Rhubarb-Pineapple Custard Pie

       Pastry for 9" two-crust pie
1¼ cups sugar
    ¼ cup flour
    ¼ teaspoon salt
    2 eggs, slightly beaten
    1 can (20 ounces) crushed pine-
       apple, drained
    4 cups rhubarb, cut into ½-inch
       pieces

Prepare pastry. Heat oven to 375°. In large bowl, mix sugar, flour and salt. Add eggs and pineapple; stir in rhubarb. Pour into pastry-lined pie pan. Cover with top crust. Seal edges and flute. Cut slits in top crust. Bake 35 to 45 minutes or until crust is deep golden brown. 6 to 8 servings.

Honey Lemon Rhubarb Pie

    4 cups cut up rhubarb
    6 tablespoons flour
    2 teaspoons grated lemon rind
4-5 drops red food coloring
1¼ cups sugar
    ¼ teaspoon salt
    ⅓ cup strained honey

Combine rhubarb, sugar, flour, salt, and rind. Mix well, blend in honey and coloring. Let stand while making crust. Line 9" pie tin with pastry. Put in rhubarb mixture. Dot with 2 tablespoons butter. Put on top crust and seal edge. Grease with milk and sprinkle with sugar. Bake in Hot oven 450° for 10 minutes. Reduce heat to 350° and bake for 35-45 minutes longer.

Rhubarb-Strawberry-Orange Pie

      3 cups rhubarb, cut in 2" pieces
      3 cups sliced fresh strawberries
 ½-¾ cup granulated sugar
  1½ tablespoons instant tapioca
      1 cup fresh orange juice
  1½ tbsp. orange marmalade (optional)
      2 teaspoons orange peel
      1 unbaked deep-dish pie shell
         Enough pie dough for lattice crust

Combine filling ingredients in large mixing bowl; let stand for 15 minutes while tapioca softens. Pour filling into pie shell. Prepare lattice strips for top crust. Bake at 400° for 20 minutes; reduce heat to 375° and bake 30 minutes longer or until rhubarb is tender. Yield: 6-8 servings.

Rhubarb Strawberry Streusel Pie

      ½ cup granulated sugar
      ½ cup brown sugar
      2 tablespoons flour
      1 egg
      ½ teaspoon vanilla
      2 cups rhubarb, chopped
      ½ cup strawberries

         Crumb topping
      ⅓ cup butter
      ½ cup brown sugar
      ¾ cup flour

Mix together granulated sugar, ½ cup brown sugar, flour, egg, vanilla, rhubarb and strawberries. Place in unbaked pie shell. In a bowl, mix together butter, brown sugar and flour; sprinkle over rhubarb mixture. Bake at 400° for 30-35 minutes.

Glazed Strawberry-Rhubarb Pie

         Pastry for 2-crust pie
  1¼ cup sugar
      ⅛ teaspoon salt
      1 cup flour
      2 cups fresh strawberries
      2 cups (1-inch pieces) rhubarb
      2 tablespoons butter or margarine
      1 tablespoon sugar

Combine 1¼ cups sugar, salt and flour. Arrange half of strawberries and rhubarb in pastry-lined pie pan. Sprinkle with half of sugar mixture. Repeat with remaining fruit and sugar mixture; dot with butter. Latticed Strawberry-Rhubarb Pie: Use a lattice top to show off the beauty of this spring pie. Bake in hot oven (425°) 40-50 minutes or until rhubarb is tender and crust is browned.

Irresistible Strawberry-Rhubarb Pie

  1⅓ cups sugar
      3 tablespoons quick cooking
         tapioca
      ¼ teaspoon salt
      ¼ teaspoon ground nutmeg
  3-4 cups rhubarb (1½" pieces)
      1 10 oz. package frozen strawberries

Combine all of above and let stand for 15 minutes. Prepare pastry for double crust pie (9 or 10 inches). Spoon fruit into unbaked crust in pan. Dot with butter. Brush outer rim of crust with water and top with crust; seal and flute. Cut several decorative vents in top crust to allow steam to escape. Sprinkle with sugar. Bake in preheated 400° oven for 35-40 minutes.

Cooked Strawberry-Rhubarb Pie

Pie crust:
1½ cups flour
½ cup lard
½ teaspoon salt
⅛-¼ teaspoon baking powder
¼ cup milk

Filling:
1½ cups sugar
3 tablespoons corn starch
½ cup water
4 cups rhubarb (cut small)
1½ cups strawberries (fresh sliced)

Mix crust ingredients well; add milk. Mix and roll out. Preheat oven to 350°. Combine water, sugar, and corn starch. Add rhubarb and cook until thickened. Boil about 2 to 3 minutes. Cool slightly, add strawberries. Pour into crust and bake 50 to 55 minutes.

Strawberry and Rhubarb Pie

2½ cups granulated sugar
¾ tablespoons cornstarch
¼ teaspoon salt
4 cups strawberries, cut in half
3 cups diced rhubarb
Pastry
2 tablespoons lemon juice
4 tablespoons butter

Mix the sugar with the cornstarch and salt. Sprinkle this mixture over the strawberries and rhubarb, and mix lightly. Line two 9" pie plates with pastry. Spoon in the fruit. Sprinkle with lemon juice, and dot with butter. Top with crust. Seal and flute edges. Freeze. Makes two pies.

To Freeze: If desired, frozen pies may be removed from pie plates. Wrap tightly in freezer wrap, and seal. Label, date, and return to freezer. Store in box to avoid damage. Otherwise, pies may be stored in pie plates. Overwrap with freezer wrap.

To Serve: Return pie to pie plate. Defrost. Cut several slits in top. Bake in a 425° oven for 30 to 40 minutes or until juices bubble up through slits. Pie may be baked from frozen state. When pie is warm enough, cut several slits in top. Add another 30 minutes to baking time. If edges begin to get too brown, cover with aluminum foil. Storage time up to 6 months.

Frozen Treats

Dinky Pops

Frozen Rhubarb Mallow

Strawberry-Rhubarb Sherbet

Frozen Rosy Mallow

Rhubarb Marshmallow Sherbet

Pineapple-Rhubarb Ice Cream

Strawberry-Rhubarb Ice Cream

Rhubarb and Orange Sorbet

Rhubarb Sherbet

Dinky Pops

3 cups rhubarb
¾ cup water
2½ cups sugar
¾ cup + 1 tablespoon light corn
    syrup
½ tablespoon butter
5 drops food coloring

Bring rhubarb and water to a boil. Lower the heat and simmer until mushy. Cool slightly. Put through blender to liquefy. Add enough water to rhubarb to make 1¾ cups. Bring to a boil. Remove from heat and add rest of ingredients. Stir until dissolved. Return to heat. Bring to a boil with a lid on then turn down the temperature to medium and remove the lid. Cook uncovered to the hard ball stage or 250° Pour into ice cube trays or popsicle forms. Freeze.

Frozen Rhubarb Mallow

3 cups diced rhubarb
2 tablespoons water
¾ cup sugar
    Dash of salt
¼ pound marshmallows
2 tablespoons lemon juice
    Rind of one lemon
1 cup cream, whipped

Heat the rhubarb and water; then add the sugar. Cook until tender. Add the salt, marshmallows, lemon juice and rind. Cool. Red coloring may be added to enhance the appearance of the dessert. Beat until fluffy, add cream and freeze until firm.

Strawberry-Rhubarb Sherbet

1 pint strawberries
2 cups cooked rhubarb
2 tablespoons lemon juice
⅛ teaspoon salt
1½ cup sugar
¾ cup heavy cream

Wash berries, hull and mash. Press rhubarb and berries through a sieve, add remaining ingredients. Pour onto a freezing tray of refrigerator and freeze without stirring, until firm. When ready to serve, scrape up thin layers with an inverted spoon and beat back and forth in tray until smooth. Serve immediately.

Frozen Rosy Mallow

1 cup flour
½ cup butter or margarine
¼ cup brown sugar, packed
½ cup nuts, chopped
4 cups rhubarb, chopped
2 cups sugar
1 3-oz. package strawberry gelatin
2 cups mini or large marshmallows
1 envelope Dream Whip, prepared

Mix flour, butter, brown sugar and nuts. Press into 8"x13" pan. Bake at 350° for 20 minutes. When cool, crumble with a fork. Reserve some crumbs for topping and pat balance into freezer-proof baking dish. Combine rhubarb and 2 cups sugar. Let stand until juice forms. Cook until tender. Add gelatin and stir until dissolved. Add marshmallows and stir. Cool until mixture starts to set. Fold in prepared Dream Whip and pour over crumbs. Top with reserved crumbs Freeze.

Rhubarb Marshmallow Sherbet

32 marshmallows (½ pound)
½ cup pineapple juice
1½ cups unsweetened rhubarb sauce
   Dash of salt
1 teaspoon lemon juice

Place marshmallows and 2 tablespoons pineapple juice in saucepan. Heat slowly, folding over and over until marshmallows are half melted. Remove from heat and continue folding until mixture is smooth and fluffy. Cool mixture slightly, then blend in rhubarb sauce, remaining pineapple juice, salt and lemon juice. Place in freezer tray and freeze, stirring several time during freezing period.

Pineapple-Rhubarb Ice Cream

2½ cups rhubarb, chopped
1 cup canned, crushed pineapple, drained
½ cup sugar
⅛ teaspoon nutmeg
⅛ teaspoon salt
3 cups evaporated milk

In a baking dish combine rhubarb, crushed pineapple, sugar, nutmeg and salt. Cover and bake in 350° oven for 1 hour. Purée baked mix in an electric blender. Cool. Add evaporated milk to purée. Churn freeze. Makes one-half gallon.

Strawberry-Rhubarb Ice Cream

5 cups rhubarb, cut up
⅛ teaspoon salt
1⅓ cups sugar
2 10 oz. packages frozen strawberries (thawed and drained)
3 cups heavy cream

Place rhubarb in baking dish and bake in 300° oven for 45 minutes. Stir every 15 minutes while baking. Cool rhubarb. Purée cooled mix and thawed drained strawberries in an electric blender. Lightly whip cream and add to purée, mixing well. Churn freeze using freezer directions. Makes one-half gallon.

Rhubarb and Orange Sorbet

2 cups diced rhubarb (about 4 stalks)
¼ cup sugar
¼ cup water
1 small, seedless orange
⅛ teaspoon almond extract

Combine rhubarb, sugar and water in pan. Cook, covered, over medium heat about 20 minutes or until rhubarb is very tender and liquid is cooked to a minimum. Cool, stirring occasionally to break up rhubarb. If orange is thick-skinned, peel it. If thin-skinned, leave skin on. Cut orange into chunks, then place in food processor and process to coarse pulp. Mix orange with rhubarb and stir in almond extract. Place mixture in ice cream machine and process according to manufacturer's directions (or transfer mixture to ice cube tray and freeze, stirring occasionally to break up large chunks of ice crystals.) Makes 2 servings.

Rhubarb Sherbet

½ cup stewed rhubarb
⅓ cup (generous) lemon juice
⅛ teaspoon salt
½ cup granulated sugar
¼ cup light corn syrup
1 cup chilled milk
1 teaspoon granulated gelatin
1 tablespoon cold water
½ cup heavy cream, whipped stiff

To the unmashed, cold, stewed rhubarb, add lemon juice, salt, sugar, and corn syrup. Blend carefully but thoroughly with the milk. Then add gelatin soaked in cold water for five minutes and dissolved over hot water. Mix well and chill. Lastly, fold in the stiffly whipped heavy cream, and freeze until mushy. Stir from bottom and sides, beat one-half minute, then continue freezing until solid but not too hard. Serve either in chilled sherbet glasses, orange cups, or cantaloupe halves which have been chilled.

Beverages

Rhubarb Beverage

Rhubarb Highball

Rhubarb Punch Float

Rhubarb Ale

Rhubarb 2-Juice Punch

Strawberry Rhubarb Frost

Rhubarb Fresh Fruit Punch

Rhubarb Slush

Rhubarb Daiquiri

Rhubarb Beverage

Cook rhubarb that has been covered with water, until tender. Strain juice. To each 4 cups juice, add ½ cup sugar and ½ cup pineapple juice. Bring to boil. Chill.

Rhubarb Highball

1 cup water
2 cups sugar
3 cups diced rhubarb

Place the sugar and water in a double boiler and heat. Add the rhubarb before the mixture boils and let simmer until the rhubarb is tender. Then force the mixture through a sieve. The resulting purée may be kept in a closed jar in the refrigerator until needed, but should not be kept more than a few days. Combine one large jigger purée and one large jigger orange juice in a glass. Stir well, add cracked ice, and fill with carbonated water. (Serves 1).

Rhubarb Punch Float

Rhubarb, fresh, cut up
Sugar
Ice milk
Carbonated beverage

Cover rhubarb with water and cook until done, then put into strainer or colander to drain off the juice. Add 1 cup sugar for each quart of juice, boil until sugar is dissolved, then cool and store until needed. Red food coloring may be added if pinker color is desired. Pour this juice into tall goblets, half full and add a scoop of ice milk. To this, add carbonated beverage, stick in a straw and enjoy.

Rhubarb Ale

1 pound rhubarb
2½ cups water
Sugar
½ cup grapefruit juice
¼ cup lemon juice

Cook 1-lb. rhubarb in 2½ cups water for about 15 minutes. When rhubarb is very soft, strain it and measure the liquid left from cooking it. Add 1 cup raw sugar for each cup liquid. Heat the liquid, stirring to dissolve sugar. Let cool. Stir in ½ cup grapefruit juice and ¼ cup lemon juice. Serve over ice cubes. Serves 4.

Rhubarb 2-Juice Punch

5 cups rhubarb (cut-up)
2 cups water
1 6-oz. can orange juice concentrate
1 (6 ounce) can pink lemonade
1½ cups sugar

Bring rhubarb and water to a boil and cook until tender (on simmer). Put through a strainer (twice maybe). Add orange juice, lemonade and sugar. Bring to a boil and cool. Can add 1 quart ginger ale or 1 quart 7-Up.

Strawberry Rhubarb Frost

½ cup rhubarb sauce
½ cup frozen strawberries, thawed
½ cup crushed pineapple
2 tablespoons lemon juice
¼ cup orange juice
2 cups crushed ice

Blend at high speed in blender until frothy. Makes 2 tall drinks.

Rhubarb Fresh Fruit Punch

8 cups diced uncooked rhubarb, not peeled
5 cups water
About 2 cups honey or sugar
6 oranges
3 lemons
A few drops of red food coloring, if needed
1 quart pale dry ginger ale, chilled

Simmer the rhubarb in the water until it's quite mushy. Strain (use a muslin jelly bag if you want it really clear), and measure the liquid into an enameled kettle. Add 1 cup sugar or honey for each 1 cup of rhubarb juice, stirring over low heat until the sweetening has dissolved. Cool. Add strained juice of oranges and lemons plus food coloring; chill. Just before serving, add ginger ale carefully. Serve over ice cubes. Makes about 3½ quarts of punch before icing. In olden days, variations of this very good cooler were made for hill country weddings where oranges and lemons were rare and ginger ale was virtually unheard of.

Rhubarb Slush

8 cups rhubarb
2 quarts water
3 cups sugar
½ cup lemon juice
1 small package strawberry gelatin
2 cups vodka (optional)

Cook rhubarb in water until tender, then strain. Add 1 small package strawberry gelatin and 2 cups vodka (optional). Freeze in ice cream pail. Scoop into glasses and add 7-Up or strawberry pop.

Rhubarb Daiquiri

1½ cups chopped rhubarb
½ cup sugar
2 tablespoons water
⅔ cup rum
3 cups ice cubes
1 tablespoon lime or lemon juice

Combine rhubarb, sugar and water in saucepan. Bring to boil. Cover and simmer 5 minutes. Let cool. Combine rhubarb mixture, rum, ice cubes and lime juice in blender. Blend until smooth. Pour into chilled cocktail glasses. Makes 4-6 servings.

Tell your neighbors, friends and family!
Here's an order form for more *Rhubarb Recipes (Second Edition)*.
Send a check or money order for $7.95 plus $2.00 shipping to:
Apple Blossom Books, P.O. Box 134, Belle Plaine, MN 56011

Ship to: _____
_____
_____
_____

Tell your neighbors, friends and family!
Here's an order form for more *Rhubarb Recipes (Second Edition)*.
Send a check or money order for $7.95 plus $2.00 shipping to:
Apple Blossom Books, P.O. Box 134, Belle Plaine, MN 56011

Ship to: _____
_____
_____
_____

Tell your neighbors, friends and family!
Here's an order form for more *Rhubarb Recipes (Second Edition)*.
Send a check or money order for $7.95 plus $2.00 shipping to:
Apple Blossom Books, P.O. Box 134, Belle Plaine, MN 56011

Ship to: _____
_____
_____
_____